This book is dedicated to the Memory of the most **Reverend Taniela Takapautolo Moala,** of the Tongan Methodist Church of New Zealand, whose help made it possible for me to come and study in New Zealand in January 1980. I graduated with a Bachelor of Science in May 1985 and Master of Science (Hons) in May 1989.

Confessions of a
SENIOR PLANT VIROLOGIST
...and Pathologist...

Second Edition

SEMISI PONE
BSc, MSc (Hons) Auckland

Note: As mentioned in my other books, quotes from Wikipedia is intended to promote the good work they are doing. Researchers can easily access individual writers and author's works, from the Wikipedia bibliography, if they wish.

Disclaimer.
All the information included in this book are available in various reports and publications of the South Pacific Commission and Member Country reports. They cannot be deemed as 'privileged' information that I received in my position as Plant Protection Advisor, Co-ordinator and Head of the Plant Protection Service, South Pacific Commission (1993-1996) (now known as Secretariat for the Pacific Community).

Note....I am writing this book in a magazine style prose for ease of reading and understanding for non-Scientific readers.

Acknowledgement

I take this opportunity to thank all my former colleagues, during my days as a Scientist in the Pacific Islands, for your support and friendship which made it such a memorable experience. I am recording the work we did together for future generations because they will want to know and learn from our experiences.

If you want to add a note or correct anything on these pages feel free to sent me a message to *rainbowenterprises7@gmail.com.*

Preface

This book is about my work as a Scientist in the Pacific Islands from June 1985-May, 1996. I thought it would be a good idea to share my thoughts with young upcoming Scientists in the Pacific Region, as well as former colleagues.

Most of the work discussed, in this book, were done in my capacity as Plant Protection Advisor (formerly Plant Protection Officer) and Co-ordinator of the Plant Protection Service, South Pacific Commission (now known as the Secretariat for the Pacific Community) from June, 1993 to May, 1996. I have also included work I did in Tonga with the Ministry of Agriculture, Fisheries and Forests and in Samoa with the University of the South Pacific, Institute for Research, Extension and Training in Agriculture, EU funded Tissue Culture Project.

I have really enjoyed the camaraderie with colleagues across the Pacific and I always look forward to visiting the islands, again, one day. I hope that everyone mentioned in this book will not mind my doing so as part of my recollections of those memorable days. I may have forgotten some details and events but as far as I know this book is a more or less complete reflection of my work in the Pacific which may not have been included in my Science oriented books PLANT PROTECTION IN THE PACIFIC series of 4 prints and 4 ebooks.

Although the titles of the book contents are about the work we did, I want to comment on aspects of them that will not appear in one of my Scientific books, but are treasured memories. I am pleased that in this day and age, I have all the resources, available on the internet, to help me finish this work.

I do apologize if anyone or former colleague felt left out from these accounts. I invite all comments and additions which can be incorporated, with acknowledgement, of all contributors, in a future rewrite of this book or one of the new editions.

As mentioned, I have already written four books and four ebooks on **Plant Protection in the Pacific**, available from amazon.com, but I find there are always more details to be added as a record for future generations to use in regional plant protection.

Best regards,

Semisi Pone BSc, MSc (Hons).
SPC Plant Protection Officer (1993)
SPC Plant Protection Advisor
(1994-1996)
Co-ordinator and Head of the SPC Plant Protection Service (1993-1996)
Chief Eexcutive, Pacific Plant Protection Organization (1994-1996)

CONTENTS

Chapters

INTRODUCTION

These records of the work I was involved in during my 10 years as a Scientist in the Pacific Islands of Tonga, Samoa and Fiji are very important to the region and its people. This is the main reason why I am taking a lot of time to write them down as accurately as possible. Science works because every Scientific discovery or process is recorded for the benefit of mankind. Later generations learn about the thoughts of Sir Isaac Newton, Albert Einstein, James Watson and Francis Crick, Charles Darwin and other famous Scientists because many people record them for future Scientists to use.

In the Pacific Islands, there are very few Scientists and it is very, very important that their work is remembered and also helpful to future generations. It is only when we stand on the shoulders of those before us that progress can be made.

Chapter 1....

The Pacific Plant Protection Organization

Robert Macfarlane, former Plant Protection Officer and my predecessor at the South Pacific Commission, had done a great job the previous 6 and a half years. He had also left me a lot of work to complete. The PPPO was one of them. The Pacific Plant Protection Organization idea had been discussed many times at national and regional level the previous 8 years and I guess that means it began with Dr Robert Ikin, the SPC Plant Protection Officer (1983-1985) before Robert Macfarlane's appointment.

It was really an attempt to formulate a Regional Plant Protection Organization (RPPO) for the Pacific Islands to represent its members at international meetings and collaboration. For example, the International Plant Protection Convention (IPPC) at the Food and

Agriculture Organization (FAO) of the United Nations organizes 2 yearly meetings of the RPPOs, at FAO HQ in Rome. The RPPO representatives from around the world attend this meeting to compare notes and share information with the FAO. As Head of the SPC PPS and Chief Executive for the PPPO, I attended the meetings at FAO HQ to represent the PPPO.

There were only 8 countries of the 27 member SPC who are members of the APPPC, in 1993, and very often the PICs, who are members of the APPPC, do not attend the APPPC meetings citing travel funding constraints. I attended the Beijing 18th APPPC meeting and also the Manila 19th APPPC meeting and it was clear to me that the PPPO is a good idea. Most Pacific Island Countries were not represented at the APPPC meetings. I was there to represent the PPPO and provide the link in information that we can use in the Pacific such as Plant Quarantine Pest Information (PQPI).

Other considerations, for the establishment of the PPPO, included the size of the Pacific region and important issues that are not included in any discussions of the FAO RPPO meetings, such as the small size of the islands and **'prior informed consent'** (pic) in terms of pesticide 'dumping', for example. The PPPO also ensure that the views of the small island States of the Pacific are heard at the highest levels like the FAO UN meetings.

Pesticide Dumping

This term is always used to refer to the continued sale, at often reduced prices, of pesticides in the Pacific Islands which have been banned in most developed countries. DDT(Dichlorodiphenyltrichloroethane) is often quoted. It was used as an insecticide for banana scab moth control in the Pacific Islands and large amounts of it were still left after the collapse of the Banana Industry in Tonga, Samoa and Fiji in the late 1980s to early 1990s. DDT was found to remain active, for a long time, in natural ecosystems and accumulate in human fatty tissue causing disease and other problems.

The RPPOs also liaise on important issues like Plant Quarantine and Global and Regional Biosecurity. It was important for the Pacific Islands to have their own organization as many pests from Asia have found their way into the Pacific such as the *banana leaf roller* and *taro leaf bligh*t. These are very serious pests which have done a lot of damage in the Pacific. The islands need the biosecurity network that will keep them informed of pest movements and quarantine dangers for both fauna and flora, thus allow for quarantine action to be taken to protect the vulnerable islands from such pests. The SPC PPS provided the PQPI through its database and computer provision which will be discussed under the SPC/EU PLANT PROTECTION IN THE PACIFIC PROJECT.

At the 8th Regional Technical Meeting on Plant Protection, Noumea, New Caledonia, 21-24 February, 1994; the importance of the PPPO was impressed upon me by colleagues from around the Pacific. I took it upon myself to push for the establishment of the organization. It was good Robert Macfarlane was there to fill me in on the details.

We got some very needed help in the form of the FAO Legal Advisor Mr Richard Stein who helped us to formulate a resolution for the establishment of the organization. The IPPC Secretariat (Dr John Hedley, Co-ordinator of the IPPC Secretariat) and FAO UN counterparts (Dr Niek van der Graff, Chief of Plant Protection, FAO) were very much interested in the progress and would like to see it through. The benefits to the region is huge, linking it to the resources and information at FAO UN and IPPC,as well as 9 other RPPOs around the world.

Mr Stein had visited me in our Suva Office to discuss the wording of the resolution which will be presented to the South Pacific Conference later that year. We also traveled to Wellington, New Zealand to discuss the resolution wording with staff of the Ministry of Foreign Affairs. Our group comprised myself, Mafaitu'uga Va'asatia Poloma Komiti, Director of Programmes from SPC and Mr Edgar Cocker from the Forum Secretariat.

The New Zealand counterparts wanted to make sure that the wording and proposal is as best as it could possibly be. We cannot afford to wait another 8 years, as the work required must proceed under the PPPO mandate.

After the discussions and improvement to the resolution wording, we returned from Wellington with a positive feeling that the resolution will be well supported at the Conference.

At the 34th South Pacific Conference of October 1994, the resolution, which has been approved by the RTMPP, PHALPS (Permanent Heads of Agriculture and Livestock Production Services) and the CRGA (Committee of Representatives of Governments and Administrations) was presented. I had requested Mr Richard Stein to come and help us with the presentation and he kindly agreed.

It was a great success for everyone who had put in some work towards its establishment in the previous eight years. The 34th South Pacific Conference approved the establishment of the PPPO under the umbrella of the South Pacific Commission in 1994. The PPPO was established as part of the SPC Plant Protection Service which will provide the secretariat for its work.

I organized the PPPO Steering Executive Committee, comprising of nine member countries from Polynesia, Melanesia, Micronesia, Australia and New Zealand to meet in Suva. We planned the first meeting of the PPPO to be held at the Tanoa Hotel, Nadi, Fiji on February 22-23, 1996.

Robert Ikin of Australia, who was the SPC Plant Protection Officer (1983-1985) was chosen as the First Chairman of the first PPPO meeting. I had suggested to Bob that he should get more involved with the PPPO especially in organizing the regional work. In addition, the Steering Executive Committee should liaise with the SPC PPS on all issues and planned activities especially with a view to developing the international status of the organization. As Head of the SPC PPS and Chief Executive of the PPPO, I attend the Technical Consultation of the RPPO's at FAO HQ, Rome, Italy, on behalf of the PPPO every two years. I often like to discuss the agenda with the Executive Committee by PEACESAT (Satellite) before I leave for Rome, for example. Such close involvement by all parties will give the organization a huge part to play in regional biosecurity. My intention was to handover all regional quarantine and biosecurity work to the Executive Committee to handle. That was the main reason why I chose the members of the Executive Committee on a geographical basis rather than expertise available. I reasoned that we already have plenty of experts available to the PPPO, the key to regional coverage and action will be the geographical location of the islands. We should have members of the Executive Committee organizing the work of the PPPO in each sub-region. That is, the sub-regions will be; 1. Polynesia 2. Melanesia 3. Micronesia. New Zealand and Australia, as extra members of the Executive Committee, can provide the expert support and funds, if required, because it would be of great economic interest to keep the Pacific Islands free of dangerous damaging pests that could enter Australia or New Zealand. That will provide a 'buffer' zone of pest free ocean space for them.

Just recently, last month (February 2019) two fruit fly species were found in fruit fly traps in Auckland. One was *Bactrocera facialis* which is only found in Tonga. It does raise the huge problem of fruit flies but also the danger to New Zealand's multi-billion dollar fruit and vegetable industries.

In addition, I was also appointed as a panel member in the Committee of Experts on Phytosanitary Measures (CEPM) at FAO HQ for a period of 7 years, from 1993 to the year 2000. I was unable to complete those two terms

because of health issues and also the
problem with job security I already
mentioned before. I was rather sad that
we could not complete the good work we
started because so many people, and my
predecessors, have dedicated so much
time to the establishment of the PPPO
because of the obvious benefits for the
region. The 34th South Pacific Conference
had approved its establishment with the
task and hope that we will complete this
work.

I was looking through the current status
of the PPPO at the FAO, during the
writing of this book, and here's the
official statement on the FAO website.

Quote....

Pacific Plant Protection Organisation

The Pacific Plant Protection Organisation (PPPO)
was founded in October 1994 by the South
Pacific Conference (now Pacific Community
Conference) at its 34th Session in Port Vila,
Vanuatu.

The Land Resources Division of the Secretariat of the Pacific Community is the Secretariat of the PPPO and runs the day-to-day affairs of the organisation. The PPPO has the responsibility of coordinating harmonization of phytosanitary measures, foster co-operation in plant protection and other phytosanitary matters among and between Members and countries and organisations outside the Pacific region. The PPPO also act for the members in developing contacts with, and where appropriate providing input into, other global and regional organisations that have authority in such matters.

PPPO is one of the Regional Plant Protection Organizations recognized by the International Plant Protection Convention and exists to provide advice on phytosanitary measures in order to facilitate trade without jeopardizing the plant health status of the importing Members and countries and in particular: 1. to ensure that the views and concerns of Pacific members are adequately taken into account in the development and implementation of global phytosanitary measure 2. assist in the development and implementation of effective and justified phytosanitary measure 3. provide a framework for regional and global co-operation in phytosanitary matters consistent with international principles for trade in plants and plant products 4. facilitate the flow of information among Members and with other regional plant protection organisations and 5. collaborate with the SPC Plant Protection Service (now as part of the Land Resources Division) on specific issues including pesticides and integrated pest management.

Members

All Members of the Pacific Community are Members of the Pacific Plant Protection Organisation. The Pacific Community consists of twenty seven (27) members including twenty two (22) Pacific Island Countries and Territories (PICTS) and 5 founding members. Pacific Island Countries and Territories Members are: American Samoa, Cook islands, Federated States of Micronesia (FSM), Fiji Islands, French Polynesia, Guam, Kiribati, Marshall Islands, Nauru, New Caledonia, Niue, Northern Mariana Islands (CNMI), Palau, Papua New Guinea (PNG), Pitcairn Islands, Samoa, Solomon Islands, Tokelau, Tonga, Tuvalu, Vanuatu, and Wallis and Futuna.

The four remaining founding countries are: Australia, France, New Zealand, and the United

States of America. The United Kingdom withdrew at the beginning of 1996 from SPC (at the time of The South Pacific Commission), rejoined in 1998 and withdrew again in January 2005.

Unquote....

The implications of the seemingly mundane meetings I had with Richard Stein, Poloma Komiti (Director of Programmes) and others before the resolution was presented to the 34th South Pacific Conference is huge with lasting influence on the quality of plant quarantine standards and trade in the region and globally. Mr Stein was an eloquent and highly knowledgeable elderly gentleman and lawyer. He impressed upon me the importance of getting the right legal framework for the PPPO. It did occur to me that the 27 member countries of SPC forms one of the most powerful groups on Earth and my job is to ensure that we do everything correctly because the PPPO is a very, very important tool for trade, quarantine, biosecurity and related work in SPC member countries.

MOU

Prior to the first meeting of the PPPO we were advised that it will be an excellent idea to have an agreement with the APPPC, for the organizations to work together. Mr Richard Stein was again of much help. After the MOU had done the usual rounds for comments, I requested approval from the Secretary General (Mr George Sokomanu) for the Director of Programmes (Mr Poloma Komiti) to sign the agreement/MOU, on behalf of SPC, with the APPPC Senior Plant Protection Officer (Professor Shen). Mr George Sokomanu, SPC Secretary General, kindly agreed.

At the first PPPO meeting in February 1996, the PPPO represented by the SPC Director of Programmes, Poloma Komiti, and the APPPC represented by the Senior Plant Protection Officer Professor Shen signed a Memorundum of Understanding

(MOU) that both organizations will work together. It formed a region of more than 50 countries in the Pacific and Asia with a very large population more than half the Earth's total inhabitants. I did realize, at the time, the enormity of this gesture. A group consisting of more than half the world's population and probably more than 60% of its wealth is a formidable force for ensuring our voice is heard in the global scheme of things, especially in biosecurity and trade.

Mr Robert Ikin, the first Chairman of the PPPO meeting, was very helpful with our discussions. He was working for AQIS (Australian Quarantine and Inspection Services) and has a wealth of experience and knowledge. I thought that he would be the right person to steer the discussions since the PPPO idea began circulating during his time as SPC Plant Protection Officer (1983-1985). I was sitting beside him at the PPPO meeting and so it was an added advantage for us to share information and comments during the general discussions. The first

PPPO meeting was also unique in that Dr CAJ Putter of FAO presented his 'Global Network' idea. The internet has just been launched and email was possible between organizations although no one knew, at the time, that the internet will take over everything like it has to-day. Dr CAJ Putter did predict that the internet will be very big in the future, but we were not on the same wave length I suppose, although I did start using email and can see why.

I had provided all 27 member representatives, at the PPPO meeting, with a computer so Dr Putter can demonstrate his network idea…and he also made some pretty accurate comments regarding the development of the internet. It was also timely to share that information with RTMPP and PACINET attendants and our Plant Protection Database which Bob Macfarlane had developed.

The RTMPP and PACINET meetings were held 'back to back' with the PPPO first meeting in February 1996 to share and reduce costs of participant travels, per diems and resources.

I was grateful for the assistance of a local Fijian friend, Mr Frank Chan, who has a bit of expertise in computers and was kind enough to provide and setup the 27 computers for the meeting. That kind of expertise was very rare in the Pacific Islands at the time. I must say that was money well spent. The SPC/EU Pacific Plant Protection Project sponsored this exercise as part of its activities. As Manager of the $NZ5 million project, I was very happy with the outcome. We also supplied the ACP countries with computers and the Plant Protection Database with Mr Chan's help. As usual there were criticism of how I did things, but I never listen to critics. They are usually the ones with no money criticizing how I spend my project money which I feel was very justified.

Trade Negotiations

One of the very important activities which I funded as part of the SPC/EU and PPPO was trade negotiations between the islands. Tonga, Samoa and Fiji were interested in selling vegetables to each other and the Forum Secretariat arranged it. The SPC/EU Pacific Plant Protection Project funded 2 Officers from Fiji Quarantine to visit Samoa and Tonga to carry out the PRA (pest risk assessment talks) after which trade in fruits and vegetables began. Tonga, for example, was able to sell watermelons to Fiji and Samoa, because they can produce large amounts, which were very popular in Fiji and Samoa, especially in the hospitality industry like hotels, motels and restaurants.

Chapter 2....

The International Plant Protection Convention

The IPPC was established in 1951 by members of the Food and Agriculture Organization (FAO) of the United Nations.

Here's an excellent description from Wikipedia;

Quote...

The International Plant Protection Convention (IPPC) is a 1951 multilateral treaty overseen by the Food and Agriculture Organization that aims to secure coordinated, effective action to prevent and to control the introduction and spread of pests of plants and plant products. The Convention extends beyond the protection of cultivated plants to the protection of natural flora and plant products. It also takes into consideration both direct and indirect damage by pests, so it includes weeds.

The Convention created a governing body consisting of each party, known as the Commission on Phytosanitary Measures, which oversees the implementation of the Convention. As of August 2017, the Convention has 183 parties, which includes 180 United Nations member states, the Cook Is, Niue and European Union. The Convention is recognized by the World Trade Organization's (WTO) Agreement on the application of Sanitary and Phytosanitary Measures (the SPS Agreement) as the only international standard setting body for plant health.

While the IPPC's primary focus is on plants and plant products moving in international trade, the Convention also covers research materials, biological control organisms, germplasm banks, containment facilities, food aid, emergency aid and anything else that can act as a vector for the spread of plant pests – for example, containers, packaging materials, soil, vehicles, vessels and machinery.

The IPPC was created by member countries of the Food and Agriculture Organization of the United Nations. The IPPC places emphasis on three core areas: international standard setting, information exchange and capacity development for the implementation of the IPPC and associated international phytosanitary standards. The Secretariat of the IPPC is housed at FAO headquarters in Rome, Italy, and is responsible for the coordination of core activities under the IPPC work program.

In recent years the Commission of Phytosanitary Measures of the IPPC has developed a strategic framework with the objectives of:

• protecting sustainable agriculture and enhancing global food security through the prevention of pest spread;

• protecting the environment, forests and biodiversity from plant pests;
• facilitating economic and trade development through the promotion of harmonized scientifically based phytosanitary measures, and:
• developing phytosanitary capacity for members to accomplish the preceding three objectives.

By focusing the Convention's efforts on these objectives, the Commission on Phytosanitary Measures of the IPPC intends to:

• Protect farmers from economically devastating pest and disease outbreaks.
• Protect the environment from the loss of species diversity.
• Protect ecosystems from the loss of viability and function as a result of pest invasions.

- Protect industries and consumers from the costs of pest control or eradication.
- Facilitate trade through International Standards that regulate the safe movements of plants and plant products.
- Protect livelihoods and food security by preventing the entry and spread of new pests of plants into a country.

Unquote...

The Co-ordinator of the IPPC, Dr John Hedley and Dr Niek van der Graff, Chief of the FAO Plant Protection Service had convened the meetings of the RPPOs and CEPM (Committee of Experts on Phytosanitary Measures) in Rome every 2 years and I was invited as the representative of the SPC-PPS and PPPO.

I must say it was an excellent opportunity for the Pacific Island countries to be represented at this level in global biosecurity. It really benefited me and my work in the Pacific region as you will read about in this book. I cannot emphasize it any better, the importance of being linked to FAO HQ in Rome and through the other RPPO's around the world.

The global movement of pests and disease was an important issue that need to be communicated to each RPPO around the world at the speed of the internet.

At the time, in February 1996, email and the internet were just in their infancy.

Many Pacific regional pest and disease problems like the *taro leaf blight* epidemic in the Samoas in 1993 and *coconut scale* outbreak in Tuvalu in 1995 were excellent examples of why we have to be vigilant on a regional and global scale. Taro is a very important crop in the Samoas akin to rice in Asia. Samoa was quoted by Dr Semisi Semisi, Head of Research, Samoa, as earning $NZ10 million from taro exports prior to the TLB epidemic in 1993. The Industry was destroyed in a matter of weeks by *taro leaf blight*. Nothing was left of the taro plant, no leaves or corms are produced.

The SPC Plant Protection Service convened the largest meetings of experts on *taro leaf blight* at the University of the South Pacific in Samoa in 1993 and University of Technology, Papua New Guinea in 1995 to put together the brightest ideas on solving the problem from all the available expertise in the world at the time.

Coconuts are very important tree crops in all of the Pacific Islands and any attack by pests or disease is taken very seriously. The SPC Plant Protection Service introduced some natural bio-control agents of the *coconut scale,* it has been using in the past, to control the coconut scale outbreak on one of the Tuvalu Islands.

Current RPPO's include;
(according to FAO, United Nations)

1. EPPO - European and Mediterranean Plant Protection Organization (EPPO)

2. NAPPO - North American Plant Protection Organization

3. APPPC - Asia-Pacific Plant Protection Commission

4. Caribbean Agricultural Health and Food Safety Agency (CAHFSA)

5. Inter-African Phytosanitary Council (IAPSC)

6. Comunidad Andina (CAN) (South America)

7. Comite de Sanidad Vegetal del Cono Sur (COSAVE) (South America)

8. Near East Plant Protection Organization (NEPPO)

9. Organismo Internacional Regional de Sanidad Agropecuaria (OIRSA)

10. Pacific Plant Protection Organization (PPPO)

> My colleagues from other RPPO's play a very important role in the informative functioning of the SPC-PPS and also the PPPO. The sharing of information of pest movements on a global scale is a top priority.

In my capacity as the Chief Executive of the PPPO, I invited the 7th meeting of RPPO's to SPC HQ in 1995 after the establishment of the PPPO by resolution of the 34th South Pacific Conference in October of 1994. That was before the opening ceremony of the new buildings and meeting room facilities at SPC. It was a world class development judging from the comments of our guests.

According to Bob Macfarlane, 'it was a great coupe' for the Pacific to bring the other RPPO's to Noumea, New Caledonia, to SPC HQ, and also to bring some awareness to the region and its pest and disease issues. I was very pleased with the meeting and the excellent cocktail reception Secretary General George Sokomanu put on for us at his residence on the first night. Our guests never expected that level of entertainment.

For the reader's information our guests at the 7th RPPO in the Noumea meeting included;

1. Dr Niek van der Graff, Chief of Plant Protection, FAO
2. Dr John Hedley, Co-ordinator, IPPC
3. Dr Bruce Hopper, Chief Executive, NAPPO
4. Dr Smith, Chief Executive, EPPO
5. Mr Ian Campbell, Chairman CEPM
6. Professor Shen, APPPC
7. Representative of Australian
8. Mr Richard Ivess, New Zealand
9. Representative of French Polynesia
10. Representative of New Caledonia
11. Representative of Wallis and Futuna
12. Seumanutafa Dr Malcolm Hazelman, SPC Agriculture Programme Manager
13. Mr Allan Allwood, SPC/FAO/UNDP Fruitfly Project
14. Mrs Fusi Taginavanua, SPC Director of Services
15. Ati George Sokomanu, Secretary General, SPC
16. Dr Brian Thistleton, SPC/EU Taro Beetle Project Teamleader
17. Observers

The PPPO has come along way since 1996, I was reading through the Draft

Report of the Third Meeting of the PPPO, held on March 19-23, 2001 in Nadi, Fiji and it was 90 pages long! I was very impressed with the detailed progress of the organization only 5-6 years since my time at SPC. It does show that the PPPO is in good hands and the work we started has progressed and improved considerably. The foundation we had laid during 1993-1996 is bearing abundant desirable fruits that will benefit the Pacific region and its island countries in the next 100 years.

Dr Mick Lloyd (the Plant Protection Advisor and Co-ordinator of the SPC-PPS, 1996-?) and his staff were doing a great job in improving and expanding the SPC-PPS projects.

Chapter 3....

Pacific Biosystematics Network

One of the SPC PPS projects was called PACINET. It was a proposal from the Commonwealth Agriculture Bureau International(CABI) . The idea was to form a Pacific loop of the CABI programme for identifying organisms called BioNET INTERNATIONAL.

This is a description of the organization from the BioNET website;

Quote...

History

Following recommendations of CABI, 1991. Concept endorsed, July 1993, at a meeting of representatives of various sub-regions of the developing world, major expert centres of developed countries, international organizations and

UN agencies and some donors, who set up *BioNET INTERNATIONAL Consultative Group (BICG)* - (BIOCON) to promote and foster the Network. Original full title *Global Network for Biosystematics of Arthropods, Nematodes and Microorganisms*; subsequent full title: *BioNET INTERNATIONAL - Global Network for Biosystematics (BI) - Global Network for Biosystematics of Invertebrates and Microorganisms.*

Aims

Promote the science and use of taxonomy, especially in the economically poorer countries of the world.

Unquote...

I thought the idea of identifying the unknown flora and fauna of the Pacific will be brilliant, especially the microorganisms. That was the main reason why I supported the proposal and used much of the SPC PPS resources to ensure we get a good result. As Manager of the $NZ5 million SPC/EU Pacific

Plant Protection Project I was in a position to fund activities that are related and complementary to my project plans. Training in Plant Protection was one of them. I thought that PACINET would be the best vehicle for our Plant Protection Training in Taxonomy and Systematics.

Professor Tecwyn Jones, who was the main force behind BioNET INTERNATIONAL at CABI had liaised with me the logistics and finance arrangement for the meeting. CABI will put in $US100,000 towards the cost. The SPC PPS will provide the Secretarial and local support of the meeting at the Tanoa Hotel, Nadi, Fiji.

I attended the CABI BioNET Global Workshop (BIGW1) at Cardiff University , Wales in 1995 to present our project statement and also listen in on other region's proposals, from around the world. It did emphasize the importance of this project to Pacific Plant Protection as identification of pests and disease is the very first step to designing any lasting control strategies. I did meet some of the representatives from other regions who were very keen on the establishment of their **loop** as well. A loop as used in this context of BioNET is known as a **'locally owned and operated partnership'**. For example, PACINET is a loop for the Pacific region.

To progress the Pacific LOOP, I proposed a meeting of the 27 member countries to discuss the proposal from CABI. Two representatives from each country and territory were invited. One from Agriculture and one from the Environment Departments. The meeting will be held at the Tanoa Hotel, back to back with the PPPO meeting and RTMPP meeting in February 1996. The SPC-PPS (Plant Protection Service) will provide the Secretarial support, organization and arrangement of the PACINET meeting.

By mid 1994, SPC Management had required all SPC International Meetings, Seminars and Workshops to be advised to them, at least, one year before the event.

As SPC is a bilingual organization (French and English), logistics for interpreters and resources need time for travel planning and transport of equipment. It may sound unbelievable but the SPC interpreters have a large amount of equipment to be transported from Noumea, New Caledonia to where-ever the meeting is held. In this case the Tanoa Hotel, Nadi, Fiji. This is a SPC requirement when some of the meeting attendants are from France or the French territories of French Polynesia, New Caledonia and Wallis and Futuna.

I had to write up all the meeting announcements for the Editors, Translators (all meeting announcements must also be translated into French for French speaking SPC member countries) and Management approval. The SPC-PPS had a total of 11 international meetings between 1993 and 1996 and it was quite an eye opening exercise as this is a very diplomatic affair. All invitations announcements from SPC are sent to the Departments of Foreign Affairs, or similar, of the receiving member countries for action. Observers are invited on an 'organization' basis.

The SPC PPS meeting announcement and proposals were all approved by the Secretary General, Ati George Sokomanu, and it was sent out to member countries and observers including the SPC/BioNET meeting. It was to be held on 12-16 February, 1996.

Although we had 3 large meetings, back to back, at the Tanoa Hotel we managed to run them smoothly with the help of the Tanoa Hotel staff. It is something that need to be recorded and acknowledged. It may help others at SPC, or other organizations, who organize large international and regional meetings. I would have welcomed such information when I was planning all those 'huge' international meetings for SPC. The SPC/BioNET meeting require 5 days and the RTMPP and PPPO meetings 2 or 3 days.

The meeting country representatives put together a work programme and budget of $US8.3 million for funding the PACINET loop. I think that is about $NZ25 million at the exchange rate of 1996. Many of the critics were scathing and called the budget 'stupid' because of the huge amount involved. I thought it rather small because it has to cover expenses of CABI as well as Environment and Agriculture Departments in 22 Island Nations but I did not want to slow down progress. We need to get the budget approved. We can always ask for more money later!

There were 54 country reps and, at least, 20 observers, from various organizations around the world, who helped with this exercise. I must say there was a lot of criticism of the meeting, programme and budget…directed at myself…but the result, after 22 years, speak for itself.

I am pleased that the project is still going strong, judging from the SPC website statement of 2010. Dr Mary Taylor, one of my colleagues, from the Tissue Culture

Project at the University of the South Pacific, IRETA, is the PACINET contact person.

I hope they will publish the list of new flora and fauna identified through the project and perhaps any trained personnel. It would be good for past colleagues to check on the progress.

Chapter 4....

Regional Technical Meeting in Plant Protection

The RTMPP is the SPC meeting of the experts in Plant Protection from each member country. There are also observers that have interests and projects in the Pacific Islands who attend the meetings every two years.

Observers usually include;

1. Australian Centre for International Agriculture Research (ACIAR)
2. Landcare Research, New Zealand (used to be known as the Department for Scientific and Industrial Research or DSIR)
3. Plant and Food Research, New Zealand
4. ORSTOM, France - French Research Institute for Development

5. IRETA, University of the South Pacific
6. Commonwealth Agriculture Bureau
7. United States AID Agency
8. United States Agriculture Department
9. Australian Aid Organization (AIDAB)
10. South Pacific Regional Environmental Programme (SREP)
11. Various Universities and other visitors depending on areas of concern.

One of the most important work of the RTMPP is to make recommendations to the PHALPS (Permanent Heads of Agriculture and Livestock Production Services) Meeting on issues of regional importance in the area of Plant Protection.The PHALPS Meeting will prioritize these issues for the CRGA (Committee of Representatives of Governments and Administrations) who will recommend them to the South Pacific Conference to be adopted as part of the SPC work programme.

For example, the establishment of the PPPO; the resolution had to go through the RTMPP, PHALPS and CRGA before being approved by the South Pacific Conference. That is after 8 years of discussions around the Pacific region at various governments and meetings.

Most SPC work programme approved items take some time to go through the SPC machinery before they are actioned. The establishment of the PPPO went through three Plant Protection Officers, a total of 12 years in work contracts, and many meetings and discussions.
The progression of the work and its complexity from Dr Robert Ikin, Mr Robert Macfarlane, myself (Semisi Pone) and Dr Mick Lloyd is quite an outstanding achievement. I should congratulate Dr Mick Lloyd and his staff for the excellent work they are doing.

The result of RTMPP, judged from the work programme and meeting reports of the first, second, third PPPO meetings is simply outstanding. The PPPO is a vehicle for the Pacific Islands to join the rest of the world in terms of Plant

Protection issues and responses. The destruction of taro in American Samoa and Samoa in 1993, by *taro leaf blight,* is one justification that is still painful to many Samoan growers.

I was impressed at the RTMPP8 when I attended as PPA (Plant Protection Adviser), Robert Macfarlane was chairing some of the sessions that he was very good at directing proceedings. I had attended the RTMPP6 in Auckland and also RTMPP7 in Suva as a country representative and I remember how I felt. I have always had problems standing in front of people to make a presentation. I was rather shy. However, when I watched Bob direct the meeting, I realize that I have to be the leader of the meeting otherwise it will be rather aimless.

Many representatives and guests complemented me on how I handled all the SPC PPS meetings. I was even surprised myself. The shy little kid is gone, replaced by a very strong willed, smart and hard working man. My wantok

Mr 'Aleki Sisifa, who later became Director of the Land Resources Division was very complementary when he attended one of our meetings as the Co-ordinator for PRAP (Pacific Regional Agriculture Programme).

I remember telling 'Aleki that I have completely lost my shyness, as a kid. I used to be part of our *'Fakame'* or White Sunday children's drama and item performance at our little church of Kapeta FWC in Nuku'alofa on the first Sunday of May every year. I was always scared of standing up, in front of the church congregation, and usually forget some of my lines of bible verses or hymns. I had developed a very bad case of 'stage fright' at a very early age and it did embarrass me on many occasions when I stumble over my words because of the affliction. Almost like a kid with a chronic stutter.

At the regional level, I meet Scientists who are experts in their field and have worked for a lifetime. Dr Douglas

Waterhouse is one example. I first saw
and heard Dr Waterhouse, when he was
70 years old, leading the discussions of
Biocontrol success in Australia, and the
Pacific Islands, at the first Biocontrol
Workshop at Vaini Research Station,
Kingdom of Tonga, 1985. Ten years later,
I was the Head of the Plant Protection
Service at SPC and I invited Dr
Waterhouse, with great pleasure, to our
Second Biocontrol Meeting held at the
Mocambo Hotel in Nadi, Fiji, 1995. Ten
years from the first meeting in Tonga. Dr
Waterhouse was 80 years old but he was
still as sharp as ever. He even asked me to
let him write up the meeting proceedings.
I was really thankful for that because our
meeting reports were accumulating and
my Information Officer cannot help.

I decided to hire Dr Graham Jackson to
help me finish the reports and book
publications of the SPC PPS. Some of
them were several years late! I guess
Doug and Graham were very aware that
our publications need to move a bit faster.
I cannot hide my delight that all our

reports were produced professionally with Doug and Graham's help.

The RTMPP9 at the Tanoa Hotel, Nadi Fiji, made me realize the huge amount of accumulating work that I was doing. It was the main reason why I hired Graham to help me. Graham had been doing a lot of the SPC publications when he was Plant Health Officer during Robert Macfarlane's tenure. His skill, experience and expertise were put to good use. After all, my $NZ5 million SPC/EU project had a large budget for publications which was considered one of the key areas that need attention. I agree with this approach 100%, because I was often referring to Graham's publications many years before for information and I realize how important they are. We now have the internet and all the information at our fingertips but it wasn't so easy to get hold of specific specialist information before the advent of the internet.

I now go through the past SPC PPS publications, and FAO publications and reflect on what could have been. The information, from all those meeting reports, are priceless. That is on reason why I am writing up all my experiences and past Plant Protection work for future generations. I hope they find them useful.

Chapter 5....

The Fruitfly Project

Fruitflies were the single biggest limiting factor to export of fruits and vegetables from the Pacific Islands, and Australia, to the lucrative market of New Zealand.

I can still remember the huge disappointment and pain of a very promising project being destroyed by biosecurity breaches in New Zealand involving fruitflies from Tonga.

When I graduated, with a Bachelor of Science, from the University of Auckland in 1985, I had this idea that I could start an agro-export business from Tonga to help my family. I rang Turner's and Growers and booked an appointment to talk to one of the executives. They said yes, somebody will meet me to discuss my idea. I got a shock when it was one of the owners, Mr Richard Turner, who met me in one of the offices and after our discussion he agreed to buy my produce from Tonga. I simply send him a telex and inform him of the produce I will send. There were no faxes or emails in those days.

I returned to Tonga in June 1985 and got a job with the Ministry of Agriculture, Fisheries and Forests as an Agriculture Officer/Plant Pathologist. I also persuaded my father to join me in borrowing some money from the Tonga Development Bank for the venture. We borrowed $4,000 from the TDB and I began the land preparation for the export crops. I planned to plant red capsicum/sweet peppers, yams and watermelons to start with. I started a nursery and sowed the capsicum from seeds bought locally. I think it was the Fua 'oe Fonua Agriculture Store that I bought my supplies of seeds, fertilizer and chemicals from. Paul and Ta Karalus were doing a great job in supplying local growers with their seeds, fertilizer and chemicals.

After the capsicum transplant, in a few weeks, the plants were growing quickly. The fruits were abundant and huge compared to most local capsicums. When the first ones were starting to turn red. I thought the quality was quite good. I estimated we will have a good first harvest so I telexed Richard Turner I will be sending some capsicum by air.

Our first harvest filled our little 1.5 tonne truck. Me and my brothers organized our first packing exercise in the kitchen on our dining table. It was the best and cleanest place for it. I haven't build a packing shed yet….and I was already thinking of building one at the farm.

We brought in the freshly woven coconut baskets full of the half ripe capsicum and did our 'quality control' and packing there. Only the capsicums with no blemish were packed. Any blemish, however small, and that fruit will be rejected. It will be sent to the local Talamahu Market in mid-town Nuku'alofa.

My first air shipment to Turner's and Growers were 38 x 3-5kg cartons of ripening green/red capsicums. It was sent from the old Fua'amotu Airport terminal which was still 'quiet' with no crowds in those days.

I got a telex back from Richard that the capsicum were 'good quality' and the auction price was good. My share of the price was $3 per carton. My very first export sale was almost $NZ100 for 38 cartons. That was a huge amount of money in those days. I was working, as a student, in Auckland, New Zealand for about $150 a week! Only 38 boxes of capsicum and I got almost 1 week wages. I thought that was great!.

Our export project was taking off. I immediately planned to build a packing house, at the farm, and buy a bigger truck. Our local capsicum/sweet pepper sale was also bringing in more than $100 a week, as well! My Mum was really keen on selling our capsicum at the local market.

She knows she is making more money than the other stall owners because of the good quality of our produce. The large mouth watering ripening capsicums were simply irresistible to the buyers at the Talamahu Market.

I estimated that our second harvest will be bigger than the first, but a huge bombshell exploded in the form of fruitfly maggots being discovered in a container of watermelons from one of the big Tongan exporters. It was all over the news on the island. It had been a wet winter in Tonga and they suggested that fruit flies were laying their eggs in the split ends of some overripe melons. The ripe melons were harvested late and the excessive amount of rainwater causes the end to split. The fruit flies lay eggs in the broken skin area of the melon which hatch in about 2 days in transit and fruit fly larvae will be rather large by the time they get to the fruit shops, in Auckland, in a week or so.

The decision from New Zealand MAF authorities was to ban all fruits and vegetable export from Tonga until further notice. I was devastated. My watermelons and yams were already on the drawing board ready to be planted, as well as expansion of the capsicum plantings. I thought we could do with $NZ300 or more from the capsicum export per week as well as up to $200 a week from the local market. That was really good money considering my wage as an Agriculture Officer/Plant Pathologist was only $TOP163, per fortnight, in 1985. Mind you, that was good pocket money for a 24 year old single man in Tonga. Most Tongan people, my age, earn that much money in 2 months rather than 2 weeks in 1985.

My plan for containers of watermelons and yams will bring in enough money to build me a brand new house, packing house and buy a new farm truck, as well as pay my brothers a full time income. I was paying my brothers an allowance a week and look forward to sharing the reward of our success and even building a house for each of them. The export ban was a huge disappointment for me and the family.

Our export project was shelved.

When I joined SPC, as the Plant Protection Officer and Coordinator of the SPC PPS, 7 years later, and visited the SPC/FAO Fruitfly Project facility at Koronivia Research Station, outside Suva, I can already appreciate the importance of Fruit fly control to facilitate trade in the Pacific Islands.

I had met Allan Allwood a few times when he visited to check the project progress in Tonga. The Fruit Fly Project had recruited a UN Volunteer to do the work in Tonga based in our Entomology Laboratory, MAFF Research at Vaini, Tongatapu. I was the Senior Plant Virologist and also responsible for the Plant Pathology Laboratory and Staff next door. I had many discussions with Pontiano, the UN Volunteer, about his work and his progress as well as Allan when he visits.

Allan Allwood, the UN/FAO expert, and his team were doing a great job. The project was being implemented by UN Volunteers, helping Allan, in 7 SPC member countries initially which was extended to include 15 countries later on. They include Fiji, Tonga, Samoa, Cook Is, Vanuatu, Solomon Is, Federated States of Micronesia, Nauru, New Caledonia, Papua New Guinea and others. The main objective was to find ways of controlling Fruitflies in a way that will allow fruit and vegetable exports, from the Pacific Islands to New Zealand, to resume. With the help of ACIAR and others, detailed study of the fruit fly biology and spread and how to control them in the field and treatment of fruits and vegetables for export.

The SPC/FAO Fruit Fly Project was assessed and proposals for its extension with a US$1 million budget was approved by UNDP. I was invited to the Extension

Workshop at the Warwick Hotel, Coral Coast, Fiji, in my capacity as Co-ordinator and Head of the SPC-PPS, to support the extension of the project because of its vital importance to the Pacific Island Regional and International Agriculture Trade. Dr Richard Drew of Australia, one of the project experts, in his speech, made me well aware of the importance of the project and its continuation.

I am pleased with the progress that the project has made over the past 22 years and how it has helped many of the Pacific Islands to restart their exports again. In Tonga's case, export to New Zealand restarted again after extensive studies and negotiations of various protocols for production, harvests and post-harvest treatments.

I read about the progress of the project in Nauru and it really highlights its importance. Children have never tasted a ripe mango because all the fruits were destroyed by fruit flies… are now able to eat ripe mangoes due to the huge success of targeted control measures by the Regional Fruit Fly Project. I think Dr Richard Drew, Allan Allwood, the Sponsors and Donors, as well as their project staff should be very proud of this achievement. ACIAR has no doubt made it all happen for the kids.

I know what it means to a child, in the islands, to eat ripe mangoes or even look at a tree heavily laden with fruits.

The success of the SPC/FAO/UNDP project in reducing Fruit fly damage in Nauru is the pinnacle of our Plant Protection work in the Pacific. If children can eat a ripe fruit where it was not available before, because of our intervention, that is something that puts Science above everything else. It becomes part of our regional legends like old, benevolent grandfather figures like Doug Waterhouse. We always remember him with fondness.

Chapter 6...

Plant Protection in Micronesia

This aim of this project aimed was to improve Plant Quarantine in the Northern Pacific. The SPC PPS placed one Plant Quarantine expert in the Northern Pacific to co-ordinate our activities there, with the view of preventing the entry of pests and disease from Asia, and other parts of the world, into the Pacific region. There are direct flights from the Northern Pacific to the Phillipines, for example...but also fishing boats from Asia coming into the region. The SPC PPS aim to improve the regional activity on biosecurity on both air and sea transport to reduce the chance of introducing a serious pest into the Pacific Islands. We consider the North Pacific Islands a weak link in regional biosecurity.

The SPC recruited Dennis Kelly from AQIS for this job. I was happy that we can get somebody from Australia because Australia has some very good links and commercial interest in keeping Asian pests out. It is right next door to Asia. I have travelled to PNG, Irian Jaya, Thailand and Malaysia via Australia and I can see their point of view. Although Australia has very good border control, in its direct flights and sea travel from Asia, the Northern Pacific Islands do not and very serious pests can enter Australia through the Pacific Islands. It would be much better to prevent pest entry into the Northern Pacific Islands, as a quarantine measure.

I travelled with Dennis Kelly, the new Quarantine Officer, up to FSM, Palau and Marshall Islands to discuss issues and introduce him to the counterparts there. We stopped over in Guam to have a look around. It occurred to me that the United States presence in their Northern Pacific territories should also make it important we co-ordinate activities. Many of the US officials dealing with agriculture in the region attend our meetings at RTMPP and

PHALPS level and share information with us on a daily and weekly basis.

The Chief Executive of NAPPO, Dr Bruce Hopper, also share information with us regarding pest movements in Canada, USA and Mexico. I also linked Dennis Kelly to our PEACESAT (Plant Protection Satellite Network) to call on us for emergencies and listen in on our weekly regional meeting.

The Northern Pacific is a huge area and Bob Macfarlane had started the project to try and improve regional plant quarantine connections which I also put much effort into realizing how vital it is to ensure biosecurity is top notch in the Northern Pacific Islands.

We had received counterpart funding from Australian AID, of $AUD 750,000, to fund the project initially and I organized for the finance requests and implementation of reports for Mr Kelly to be coordinated with me and the Finance Manager in Noumea, New Caledonia.

John Roach, our Finance Manager in Noumea, is always on the ball and I was pleased that our finance is in good hands. Anything I request is attended to very quickly and efficiently.

It must be said that regional co-ordination of NPPO activity is a good way to improve regional service and biosecurity measures.

Chapter 7....

The Taro Beetle Project

Taro beetle (*Papuana sp, Papuana uninodis*) is a huge problem of *Colocasia esculenta* taro in Fiji, Kiribati, New Caledonia, Vanuatu, Solomon Islands and Papua New Guinea. I visited the Taro Beetle Project and attended the review led by the legendary Dr Doug Waterhouse. Doug was already in his seventies at the time. The project was trialing various bio-control agents and other methods of controlling the taro beetle.

Dr Brian Thistleton, Dr Billy Theunis, 'Ioane 'Aloali'i and Roy Masamdu were doing some really good work and progress has been amazing.

I was invited to the Marie Curie Symposium in French Polynesia and I asked Brian, the Team Leader, if I can take his project presentation to showcase it as my 'poster' at the symposium. It was one of the popular posters judging from the number of people viewing it when I am around. Although I had many projects myself, with the Pacific Plant Protection Project, I thought that taro beetle damage is something that everyone should see, in French Polynesia, to prevent it spreading any further than its distribution at the time.

I wanted to create some awareness of the taro beetle problem because it could really destroy efforts to cultivate taro in other parts of the region. Even though the corms are large and everything about taste, appearance and size are perfect but the large holes caused by the beetle negates all its good qualities. Taro is an important staple in most countries of Polynesia, Melanesia and Micronesia and at the time it was being attacked by two formidable pests. First *taro leaf blight* in American Samoa and Samoa and *taro beetle* in PNG, Solomon Is, Kiribati, Fiji, New Caledonia and Vanuatu.

While in the Solomon Islands, I had a chance to visit the war memorial park with all the broken machinery of war. Planes, tanks, guns and other memorabilia. *Taro leaf blight* is reputed to have been introduced into the Pacific from Asia by the Japanese army who grew taro in the Pacific to feed their soldiers. It did impress upon me the naked horror of war and its legacy which is currently waging another war against taro growers of the Samoas.

DID THE ALLIES INTRODUCE TLB?

It did occur to me that the Allies may have introduced Taro Leaf Blight to destroy the Japanese taro plantations and cause a food supply problem for their soldiers. But judging from the large amount of movements of the machinery of war between the islands....the spores of the fungus could easily hitch a ride.

After the review everybody, who attended, agreed that it is very important to continue the project and the efforts to find a solution to this very damaging pest. Dr Thistleton, did an excellent job making sure everyone is informed about their very important work.

I had dinner with Dr Thistleton and Ruth Liloqula, his wife, during my visit and we did made some small talk about the project and situation in Honiara. Ruth is a very out spoken Lady on many issues in the Pacific Islands and her concern for the welfare of her people was very obvious.

It is one of the sticking points that I have never talked about in front of Pacific Islanders, but it seems that in almost all the islands the potential for making a good income of just about everything is there. You just need to start something and keep it going until you succeed. This is also true of agriculture. If we can select a few examples, the squash exports from Tonga, New Caledonia and Vanuatu is a huge success. Watermelon export and root export from Tonga.

I discussed the cassava production with various Solomon Islanders and it appears that is something that need looking at but never proceeded further than talk. Is it the pests and disease? Or just the ability of the locals to put thoughts into action? Getting an alternative source of income while taro is being sorted.

I know that is a very offensive statement to many locals every where in the Pacific Is., but it is true. Very often it is not the islanders fault but the Governments and their politicians. Like most new projects in the Pacific Islands, it is usually expatriates who come in with some cash and ideas that start new local business. Examples include Squash Pumpkin export in Tonga, New Caledonia, Vanuatu and Vanilla Export in Tonga.

Chapter 8....

The SPC/German Biological Control Project

I was working as a Plant Pathologist at MAFF, Tonga when the first ACIAR/GTZ/MAFF Biological Control Workshop was held at Vaini Research Station, 17-25 October, 1985. Although I studied biological control principles as part of my Bachelor of Science degree, at Auckland University, it did not prepare me for the obvious importance and international network of Scientists active in the field in the Pacific. That was the first time I heard Dr Douglas Waterhouse speak and I could tell, immediately, he is one of the prominent Scientists on the subject, in the Pacific. Doug was already 70 years old at the time and was quite popular with everyone. There were a large number of Scientists present, and country representatives from all over the Pacific Islands.

Topics like the importance of controlling the Diamond Back Moth came to prominence and bio-control became the war weapon of choice against pests in the Pacific.

My auntie, Mrs Lu'isa Kefu, was doing the secretarial work and it did dawn upon me that this is a bit more than the normal workshop, judging from the importance she put on it. All visitors were to be treated as 'ambassadors', from her viewpoint. It was the first time I had come across an SPC meeting and I met some country reps who will be my counterparts at SPC, later when I became the SPC Plant Protection Officer.

We held the Second Biological Control Workshop at the Mocambo Hotel, Nadi, Fiji in October, 1995; exactly 10 years from the first meeting. Dr Waterhouse was about 80 years old at the time, but he still volunteered to write up the meeting report. Doug knew that I was up to my ears in work organizing eleven meetings, seven projects, attend international meetings and commitments and supervise staff and I have no time to write up the reports. We did have Information Officers, who I tasked with writing our meeting reports, but they were not in a position to help. The expertise of Dr Waterhouse is legendary and we thought it a great advantage.

Dr Paul Ferrar had been a very strong and active supporter of our work at SPC. I had worked with Paul, Dr Richard Davis, Professor John Brown and ACIAR in Tonga, helping them with their Kava Dieback Project. Paul had sponsored one of our resource persons, Professor Carlos Klein-Koch, our former Biocontrol expert for the project to return from Chile and give us a report of the bio-control scene in South America. Carlos had joined a University in Chile as a Professor and he was happy to give us a presentation of what the biocontrol situation in Chile looks like.

The main aim of the 1995 bio-control meeting was to review what had been done since the meeting in Tonga 10 years before. What has the Pacific achieved or improved in the field of Biological Control? As usual, all member countries and territories were represented…with reports of great success achieved.

The SPC/GTZ Biocontrol project focused their activities around the Integrated Pest Management of Diamond Back Moth, *Plutella xylostella*. The IPM programme included the use of biocontrol agents and minimal pesticide use to improve production.The IPM programme was carried out in several countries including Tonga, Cook is, Samoa, Fiji, PNG, Vanuatu and Solomon Is. I have been to some of the trials in Fiji, Cook Is, Tonga and PNG and it appears to be working well.

Another target species was *Lantana camara* with bio-agent *Uroplata girardi* and two other agents. Tonga was one success story where *Lantana* covered large areas in the 1970s but was almost all gone by the 1990s thanks to the

Tonga-German Plant Protection Project and later collaboration with the SPC/German Biocontrol Project, Dr Jurgen Schaeffer and Dr Carlos Klein-Koch. Dr Dirk Stechmann, Team Leader of the Tonga German Plant Protection Project was very active and involved in the First Biocontrol Workshop and the release of several bio-agents in Tonga including the control of *Lantana camara*.

Dr Carlos Klein-Koch, who was the Entomologist and Bio-control counterpart in the SC/German Biocontrol project left in 1994 and it was imperative to hire someone as soon as possible because of the large number of bio-control releases going on. Although our technician, Mr Humesh Kumar was able and trained to carry out the work, he was not in a position to do much of the member country releases and NPPO (National Plant Protection Organization) counterpart negotiations concerning regional biocontrol work. I was pleased when we finally got Mr Albert Peters on board as the SPC/EU Biocontrol Officer in 1995.

The SPC/German Biocontrol Project was also involved in training and similar activities in the area of biological control.

The project recruited the services of Mr Rupeni Tamanikaiyaroi of Fiji to carry out some of its work in the west of Viti Levu. I had known Rupeni from USP, when I was the 'Fellow in Tissue Culture' (April 1992- April 1993); he was a teacher there. I was hoping that we can train Rupeni to take over this role as it appears that the Pacific Islands really need local highly trained staff.

We also recruited another technician to help Rupeni with the IPM work at Nadi. The SPC/German Biocontrol Project was also heavily involved in extension training having done 2 workshops on the subject. The first one in Tonga at the Dateline Hotel in 1993 and the second in Fiji at the Sonaisali Resort in 1995.

It is interesting to note that many successes have occurred in the Pacific in the area of bio-control in the last 100 years but there are still no trained NPPO staff on it. This is my observation with NPPOs, they will always rely on expatriates to come and initiate the work or project. In some countries, funding problems are always the reasons cited for lack of Scientific work in agriculture....but then...it is also the same reason given for lack of development and infrastructure, employment and exports.

In fact, lack of funds is the main reason for the under development, of everything, in the Pacific according to most NPPOs.

Chapter 9....

The SPC/EU Pacific Plant Protection Project

Of the seven projects under the SPC PPS, I helped co-ordinate the others and managed the SPC/EU Pacific Plant Protection Project. As Manager, I organize the programmes and budget for activities every year for approval by SPC, Forum Secretariat and the EU Delegation. The European Union funded the project for ACP countries and territories, which meant only Fiji, Kiribati, Papua New Guinea, Solomon Is, Tonga, Tuvalu, Samoa, Vanuatu were included with the 3 French Territories of French Polynesia, New Caledonia and Wallis and Futuna. New Zealand ODA were asked to fund the three New Zealand Territtories of Cook Is, Niue and Tokelau and the US Territories of FSM, Guam, Marshall Is and American Samoa were, hopefully, to be funded from US contributions.

It was very complicated wading through the various issues of territories and funds. The European Union Funds for example can only be used in ACP countries and Territories.

During the RTMPP, PPPO and PACINET meetings, we supplied all the participants with computers to show them how to use the Plant Protection Database which Bob Macfarlane had put together. The idea was, that after the workshop everyone will go home with a new computer and a copy of the database which will improve Plant Quarantine Risk Assessments in the Pacific Islands. The only problem was I only had funds for the ACP countries and French Territories and we did provide them with computers and copies of the database. The other country representatives only went home with copies of the database and no computer! I really felt bad about that but it was just too late to get all the funds needed from the various donors for the February 1996 workshop. We had made the requests but there was nothing coming back.

However, I was confident of getting funds for supplying a computer to all the countries who require one. Some countries may already have a computerized Quarantine Service.

Our project ran a small quarantine information and training workshop with assistance from the Media and Graphics Department for the ACP countries. We produced some videos to help the taro leaf blight training in Samoa and various publications and printed T-Shirts for use by the NPPOs.

The project also provided funds for a Bio-Control Officer, Information Officer, Tissue Culture and Bio-control Technicians and a Project Secretary/Personal Assistant.

The Bio-control and Tissue Culture laboratories were funded by the project. I was in charge of both laboratories with the Technicians in place until the Bio-control Officer took over. The Tissue Culture lab remained under my direct supervision. There was a Bio-control expert with the SPC/German Biocontrol Project, Dr Carlos Klein-Koch, but he left after a year to take up a Professorship in Chile....and the SPC/EU Biocontrol Officer, Mr Albert Peters, did not join us until the second or third year.

We supplied beneficial insects to island countries with pest outbreaks and also tissue cultured elite crop germplasm for research and hurricane recovery. Although the funds were EU funds for ACP intervention, this service was available to all island countries. From memory, we had supplied quite a large number of beneficial insects and plantlets that were in the 1,000s during my time as the Head of the SPC PPS.

One of the most important aspects of the project was Plant Quarantine. As a member of the IPPC/FAO Expert Committee on Phytosanitary Measures and Acting Chief Executive of the PPPO, I was well placed to connect the SPC and PPPO to FAO and the RPPO meeting in Rome as well as the Global Exchange of information on Pest Movement. The partner project in Micronesia with Mr Kelly was also under my supervision to ensure that all our efforts of 'Regional Biosecurity' is well co-ordinated. I should briefly mention that the countries of Micronesia were also included in our Quarantine Training and Materials Production workshop for ACP countries funded from the Plant Protection in Micronesia project.

Chapter 10….

Plant Tissue Culture

Plant Tissue Culture is one of the elite tools that we use in Pacific Agriculture to help with Plant Protection because;

1. Tissue Cultured plantlets are free of disease and micro-organisms and can be used for germplasm exchange between countries and islands of the Pacific. It avoids the unwitting transmission and introduction of pests and disease into new areas or countries where they are not present. It is the new introduction of disease into a new area or island where it is not present which often spark epidemics like the *taro leaf blight* in the Samoas in 1993.
2. Tissue Cultured plantlets are small and can be air freighted into islands around the Pacific at low cost within 24-48 hours.
3. New improved varieties of crops can be made available for crop research and

production in island member countries of USP and SPC.

4. Tissue culture can also be used to mass produce disease free improved varieties for commercial ventures in the Pacific region. They can also be imported in places where Tissue Culture facilities are not available.

The Tissue Culture Facility at the Vaini Research Station, MAFF, Tonga was locally build to assist with the Banana Research work in collaboration with ACIAR. Large numbers of banana varieties were imported from Australia for field planting and multiplication in the local laboratory. Mr Tevita Holo and Mrs Paelata Nai were in charge of this project with some really good results despite the discontinuation of banana export.

I was involved in the USP/EU Tissue Culture Project at the Institute for Research, Extension and Training in Agriculture at the University of the South Pacific from April, 1992 to April, 1993. Our work involved the maintenance of the popular Pacific crop varieties in culture storage. I also did some research on many aspects of our work which is published in PLANT PROTECTION IN THE PACIFIC 3, Tissue Culture. The facility at IRETA/USP is world class thanks to the efforts of Dr Mary Taylor, Mr Brian Smith and the Pacific Regional Agriculture Programme. I think that the USP member countries should make use of its potential. I was in charge of supplying germplasm to USP member countries on request. We did send out a large number of plantlets to help member countries with their various projects.

The IRETA/USP germplasm collection included more than 200 taro accessions, more than 100 sweet potato accessions, some cassava, banana, and yam varieties and also vanilla.

Chapter 11....

Squash Exports from Tonga

One Friday night in 1985, I met a former 'mate' from Auckland at Joe's Hotel in Nuku'alofa. Josh Liava'a is one of those people that has a lot of charm and he used to help and coach our Tongan university student rugby league team in 1981. The Tongan Students at the University got together and formed a team, under the City Newton Club, to join the Rugby League competition. None of us had played rugby league before and Josh helped us understand how the game is played.

A former Policeman turned businessman, he was in Tonga with some New Zealand business people to promote a project to plant squash pumpkin for export to Japan. New Zealand is one of the big exporters of squash pumpkin to Japan.

Josh said, the aim is to plant them during the cool months. We had a chat about their project and they did get some growers to start with, but it was not successful as sometimes there can be 'droughts' or long periods with no rain in the middle of the year. I did warn Josh about the 'dry period' which sometimes causes crop failure.

Prince Mailefihi Tuku'aho took on the project in 1987 and promoted it with help from MAFF research and he and his group made some money from it. Growers were inspired to grow more and by 1991 there were, at least, 5 groups exporting squash pumpkin to Japan.

In the case of Tonga during the Squash Export years, 1991 was an explosive year for the virus. It was probably why growers and researchers noticed. Dr Viliami Manu and Dr Siua Halavatau, Soil Scientists with MAFF Research, were conducting trials of various fertilizer doses at the Tufumahina King's Estate when the virus damaged a large number of their squash pumpkin plants on the trial plot. Dr Manu asked me, as the Senior Plant Virologist, to look into it and I conducted surveys to determine the spread and severity of the virus, if any, around plantations in Tongatapu. It was quite a shock to find many plantations of 2-4 acres in size which were severely damaged with large areas of the crop appearing yellow because of the virus infection.

Other reports of a virus disease attacking squash in various areas on Tongatapu prompted me to expand the survey. It was certainly a new disease, no one had seen it before on their squash pumpkins. After a search of the literature and making a list of possible causes, I narrowed down the list to about 4 viruses; 1. Watermelon Mosaic Virus (WMV) 2. Zucchini Yellow Mosaic Virus (ZYMV) 3. Cucumber Mosaic Virus (CMV) 4. Watermelon Mosaic Virus II (WMVII). Our MAFF Research laboratory had been set up, by myself, for virus testing using ELISA when I was working with the Vanilla Necrosis Potyvirus. I ordered antiserum from Sigma USA for testing against these 4 viruses. When the antiserum arrived in Tonga, I did a survey of plantations around Tongatapu aiming to cover as much of the island as possible with more than 200 samples collected.

The result was not surprising at all. More than 90% of the samples tested were positive to ZYMV. ZYMV is the most widely distributed destructive virus of cucurbits in the Pacific Islands and worldwide causing significant damage to crops and economic losses everywhere according to reports in the literature.

I immediately designed control measures and we started giving advice, on the radio and newspapers, to export squash growers on how to manage the virus disease. We also ran some training workshops with advisory staff of how to advice growers on ZYMV control measures. That was an ongoing programme. We also started trials on how to control aphids using reflective mulches and barriers.

Mr Tevita Holo, Principal Plant Pathologist and myself Senior Plant Virologist designed the trials. We decided to investigate the effect of 1. dry grass 2. white plastic 3. aluminium foil and 4. bare ground on aphid migration and movement around the squash pumpkin mounds when plants are still emerging until they start flowering and fruiting. We used a yellow paint on the top 5 cm of wood poles about 2 feet high and 2cm by 2cm square, covered in grease as our trap to catch the aphids as they fly into the mound where the young squash pumpkin plants are growing.

It is well known that aphids are the carriers of the ZYMV and they are also attracted to yellow colors.

From the beginning, we noticed that our traps were catching the most aphids on the side of the prevailing wind, suggesting that wind direction plays an important role on aphid migration. Our studies of aphids, at MAFF Research, especially Dr Dirk Stechmann's study on banana aphid *Pentalonia nigronervosa* suggest that aphids start developing wings and migrating when there is an 'overpopulation' of the host plant. That is the dangerous phase because they tend to fly away from a usually infected ratoon banana plant and infect any young plants nearby.

In the case of squash pumpkin, the most important of the eight aphid species known in Tonga is *Aphis gosypii*, also known as the 'Cotton or watermelon aphid' but it is found in almost all plants and weeds, in large numbers. It is also the species most commonly caught in our traps.

All the reflective mulches were significantly better than the bare ground in 'repelling aphids' judging from the numbers caught. It has been known for some time , in many international studies, that reflective mulches repel aphids probably due to the light wavelengths reflected into the atmosphere which aphids avoid.

Although barriers like corn were used to 'trap' the aphids as they fly into the squash plantation, we did not measure or collect data of its effect on disease incidence and spread but it was observable in the field, that disease incidence appear to be less when corn barriers are used. The harvest and sale of the corn was also an economic benefit, including an additional food source for the grower's family.

Aphids are known to carry virus particles on their proboscis or feeding tube, especially potyviruses. ZYMV is a member of the potyvirus group and it is transmitted in a 'non-persistent manner'. The idea of using corn, is not only to trap the aphids, but also to remove the viruses from the proboscis as they 'probe' on the corn leaves. Aphids normally start probing plants they land on as a kind of 'tasting' for feeding purposes. The theory is that they 'lose' their load of aphids after probing and feeding on the corn, which is not affected by the virus.

The other benefit is the 'non-persistence' of the virus, which becomes non-infective in a matter of minutes.

The ZYMV control strategies advised to the Tongan Squash Export growers were;

1. Select an area that has no cucurbit plants nearby including watermelon, squash, cucumber, zucchini, buttercups and any other member of the Cucurbitaceae which can host the ZYMV. If there are infected plants near the new squash plantation, ZYMV can spread to the new plantation and totally wipe it out.
2. Plow or turn over old cucurbit plantations and remove all weeds that host aphids especially succulent, herbaceous weed like *Sonchus olearacea* and others like it from the immediate vicinity of the plantation.
3. Ensure no weeds are present inside the plantation that can host aphids in large numbers.
4. Inspect young squash plantations regularly , especially around the border, and remove any plants that show ZYMV

disease symptoms as soon as the 'yellowing of veins and leaf lamina' can be detected. Spray the infected plants first , with insecticide, to kill any aphids on it, then pull it out after a period of time, for example 30 minutes. It may be a good idea to collect those infected plants in a plastic bag and bury or burn them elsewhere if possible. If they are left on the field they may become a source of ZYMV infection for aphids that land and probe on those uprooted plants. It may take a day for those plants to dry out or even longer during rainy periods.

Aphids can also fly in with the ZYMV virus particles on their proboscis, from neighboring infected plantations and land on border plants, infecting them during probing and feeding. Replace any young infected squash plants from a potted nursery for that purpose.

Although ZYMV, like all potyviruses, is transmitted in a 'non-persistent manner', which means the virus particles do not survive on the proboscis for longer than a few minutes, it is still enough time for the aphid to infect several plants which will become a source of inoculum for the rest of the plantation. From field observations, ZYMV can spread in Tongatapu squash pumpkin plantations and totally destroy it in a matter of days. However, even though many plantations were destroyed by the virus in 1991, Tonga was still able to export 22,000 tonnes of squash pumpkin to Japan after the ZYMV control measures were put into place.

5. Use systemic insectide spray regularly, two weekly for example, on the border of the plantation to kill any aphid colonies there that may become vectors for ZYMV within the plantation.   Use systemic insecticide for aphid control during normal spray programme against fungal disease and insects.

Systemic insecticides enter the plant and remain active for 2 weeks. It is much more advantageous than 'contact insecticides' which remain on the surface

of the plant and can be washed off by
rain.

Chapter 12....

Kava Dieback

I first observed kava dieback on several kava plants my brother planted on our property at Kapeta, Nuku'alofa. I was in my teens and I can still remember the crinkling and yellow/orange spots on the young leaves. Then after a few weeks the whole plant dies back from the tip and rots giving off a terrible smell. I did learn later, when I started drinking kava, that it was the typical smell of kava mixed with the rotting smell of the stem and leaf tissue. There was continuous growth from the base that looks healthy but when the shoots are about two to three feet high, the yellow/orange spots and crinkling appeared and the dieback repeats. After several repeats the base just rots with no more new growth.

That was in the mid 1970s. I was a student at Tonga High School and it did not occur to me that it may be a huge problem for kava growers.

By 1989, after graduating with a Master of Science (Hons) from Auckland University, my work on the Vanilla Necrosis Potyvirus also highlighted the dieback problem of kava. ACIAR and MAFF began a collaborative project on kava dieback, soon after, with a PhD student, Richard Davis, based at our Plant Pathology Laboratory, MAFF Vaini Research Station.

Kava Dieback History

There has been much research on Kava Dieback in the previous 20 years but there has been no satisfactory result.

Although I have been aware of the kava disease, I did not do any work on it since I began working at the MAFF Vaini Research Station in June, 1985. Richard worked very hard on the project and had discussions with me on several occasions of his progress. After about a year he said he has exhausted all avenues and there is still no result. He had looked at the possibility of fungi, bacteria and mycoplasma as causal organisms of kava dieback and there was nothing, no result. All his experiments were negative, meaning none of them can possibly be the causal organism.

One day Richard asked me to have a look at a plantation close to the Research Station. I got a shock, it was exactly the symptoms which I had seen on our kava plants about a decade and a half before. The yellow/orange spots on the young leaf lamina and veins, the rotting stem, basal shoot growth and dieback of some of the stems. It occurred to me immediately that it must be a virus. Fungi, bacteria and mycoplasmas will not cause veinal cholorosis of kava leaves, especially when accompanied by leaf buckering, distortion and crinkling of young kava leaves. Only viruses cause those types of symptoms…together. Fungi and bacteria can cause leaf necrosis and yellowing without distortion but usually of older leaves. Mycoplasmas can cause yellowing but not distortion and necrosis of young leaves.

I returned to the Research Station and told Richard that he may be dealing with a virus and we should send some symptomatic leaves to Australia for checking on the electron microscope for virus particles.

I attended the project review to listen in and also discuss the project progress. Professor John Brown of the Australian National University, Richard's supervisor, and Dr Paul Ferrar of ACIAR were there. I suggested to them that we should try electron microscopy and search the sap of symptomatic kava leaves for virus particles. Richard sent some samples to Australia and within a few weeks we got results! They found virus particles that looks like Cucumber Mosaic Virus.

That was a huge success from Richard's point of view. After one year of hard work with no success, the result was handed to him on a plate.

I was working on Vanilla Necrosis Potyvirus which was destroying many vanilla plantations on Tongatapu and Vava'u, at the time. I had developed an ELISA (Enzyme Linked Immuno-Sorbent Assay) test to VNPV and also set up ELISA test equipment at the Plant Pathology Lab, Vaini Research Station.

After the break through with electron microscopy, I decided to order some Cucumber Mosaic Virus (CMV) antiserum from Sigma USA to develop an ELISA test for CMV on kava. Once the antiserum arrived, I tested some local symptomatic kava plants, using ELISA, and they were all positive to CMV! We planned a survey of kava plantations in Vava'u, Ha'apai and Tongatapu and tested all the collected symptomatic leaf material for CMV presence using ELISA. All the symptomatic leaves from Vava'u, Ha'apai and Tongatapu were positive to CMV. I can remember that we tested more than 100 samples…and all were positive! It was the best evidence that we can get that CMV is closely associated with the symptoms and kava dieback.

In March 1992, I joined the University of the South Pacific IRETA Tissue Culture Project. Richard Davis and John Brown continued with the project and proved beyond reasonable doubt that CMV is the causal organism of kava dieback. The results were published in the joint authored paper;

Davis, R.I., Brown, J.F. and Pone, S.P. 1996. Causal relationship between cucumber mosaic cucumovirus and kava dieback in the South Pacific. Plant Diseasc 80: 194-198.

I also include some of the work I did on kava in my book, PLANT PROTECTION IN THE PACIFIC available from amazon.com.

An ebook on how to control kava dieback in the Pacific Islands is also available from amazon.com

Chapter 13....

Vanilla viruses

The French Vanilla and Spices Project Team Leader, Mr Stephan Sorin, asked me one day to have a look at a plant on their trial vanilla plot. It had symptoms I have never seen before. The young leaves were buckered with sunken chlorotic lesions, necrosis on parts of the stem and stunting of its growth. All the other vànilla plants have grown, at least ten times, bigger. It was obviously affected by something.

I took some of the necrotic stem pieces, and asked our Laboratory technician Mrs Paelata Vi Nai to try isolating any fungal or bacterial infection from it. After several days she says there is still no growth! I had a look and it was obvious there is no fungus or bacteria in the blackened rotting stem. Normally, there will be fungal or bacterial growth from the necrotic tissue.

I contacted Dr Mike Pearson of the School of Biological Science, University of Auckland who gave me a New Zealand MAFF Plant Quarantine permit to send him some samples. He reported back that he found 'flexuous, filamentous virus particles which appear to be potyviruses' in the symptomatic sample I sent. I discussed the results of our investigation with Dr Dirk Stechmann, Teamleader of the Tonga-German Plant Protection Project. He agreed to fund Dr Pearson to visit the Kingdom and do a survey, with us, for a better idea of virus presence. Stephan was also very pleased with the results, we obtained so quickly.

We visited several plantations in Tonga and collected 30 samples which Dr Pearson took with him for further electron microscopy examination at Auckland University.

The samples were of three categories according to 1. severe symptoms 2.

mottles 3. no symptoms. The results were reported in;

Viruses of Vanilla in the Kingdom of Tonga, Pearson, MN and Pone, SP Australian Plant Pathology, September 1988, Volume 17, Issue 3, pp 59-60.

Three viruses were found by electron microscopy examination. They were Cymbidium mosaic virus (CYMV), Odontoglossum Ringspot Virus (ORSV) and a potyvirus. CyMV and ORSV are the most common viruses of cultivated orchids around the world and they normally do not cause any major problems other than the mottling of leaves, ring spots, and the occasional necrotic spot. Most domestic orchid growers do not screen their orchids for CyMV or ORSV and so they are continually spread by planting material throughout New Zealand. It appears there is a need for a screening programme to get rid of these viruses.

The vanilla potyvirus was unknown and it was very important to identify and characterize the virus before any control measures can be designed. Our survey does associate the severe symptoms with the potyvirus.

The Tonga-German Plant Protection Project included a lot of scholarships for Bachelor, Master and PhD programmes. I was given one of the MSc scholarships to do research at the University of Auckland to try and identify the potyvirus.

I began the programme in 1986 and by December 1988 we have identified and characterized the virus. It was an uncited potyvirus. We gave it the name Vanilla Necrosis Potyvirus. This work is reported in my MSc Thesis;

An investigation of three viruses of *Vanilla fragrans* (Salisb) Ames from the Kingdom of Tonga.

Other work done by myself, Dr MN Pearson, Dr Lia Liefting and Dr Allan Brunt of the United Kingdom are reported in these scientific papers;

1. Some Hosts and Properties of a Potyvirus Infecting Vanilla fragrans (Orchidaceae) in the Kingdom of Tonga. M. N. Pearson, A. A. Brunt, S. P. Pone. Journal of Phytopathology, Vol 128, Issue 1, January 1990, 46-54

2. Lia Liefting, Michael Pearson and Semisi Pone, The Isolation and Evaluation of Two Naturally Occurring Mild Strains of Vanilla Necrosis Potyvirus for Control by Cross-Protection, *Journal of Phytopathology*, **136**, 1, (9), (1992).

3. M. N. PEARSON, G. V. H. JACKSON, S. P. PONE and R. L. J. HOWITT, Vanilla viruses in the South Pacific, *Plant Pathology*, **42**, 1, (127-131), (2007).

More extensive work done my myself and other details are reported in my series of books on plant protection;

1. Plant Protection in the Pacific
2. Plant Protection in the Pacific 2
3. Plant Protection in the Pacific 3, Tissue Culture
4. Plant Protection in the Pacific 4

These books are available from amazon.com.

There are more books to be included in this series including my work on the serology of the Vanilla Necrosis

Potyvirus (VNPV), Potyvirus Group Test (from Sigma USA) and development of the Enzyme Linked Immuno-Sorbent Assay to VNPV.

The characterization of the VNPV is included in Plant Protection in the Pacific 2, including epidemiology work which suggest that VNPV can spread very quickly within a plantation not only along the rows but also 'clustering' suggesting that growers do transmit the virus during normal plantation work like pruning, flower initiation, pollination, harvesting and taking of cuttings which can transmit sap from plant to plant at a significant level of $P=0.05$.

The 'clustering' effect occurs when there is an agent or vector that spreads the virus from plant to plant within the plantation. It follows that an infection will significantly increase the chance of the neighboring plant being infected. This conclusion was significant at $P=0.05$. In one of the plantations where clustering was detected, a large number of aphids were also observed during the recording of the data during an 8 months period. The aphids, presumably *Aphis gossypii* which is the most common aphid in Tonga, were populating a common weed, *Sonchus oleraceaus* which grows very quickly in most plantations in Tonga.

Aphids are known to develop wings and 'migrate' when the host is overpopulated. Dr Dirk Stechmann (Entomologist and Teamleader of the Tonga-German Plant Protection Project) had made the comment that this is the case with Banana Bunchy Top Virus being spread by *Pentalonia nigronervosa*.

Aphid 'migration' flights is the most significant in terms of virus transmission. Aphids tend to land and probe or feed on many plants thus transmitting the virus on their proboscis from plant to plant. One of the most important pieces of information about characterization and identification of a virus is whether it is transmitted by viruses in a persistent or non-persistent manner.

In the case of VNPV, it is known to be a potyvirus which is transmitted in a non-persistent manner. This means that VNPV is transmitted from plant to plant, within a vanilla plantation, by aphids within a matter of minutes. It does pose a very serious problem because it means that the whole plantation can be infected by VNPV within a short period of time depending on the number of aphids present, flight/migration patterns. For example, if there are large numbers of aphids, within the plantation, migrating or seeking new hosts because the current one is overpopulated, then we will see a spike or sharp increase in infected vanilla plants or plants showing symptoms.

Surveys of young plantations less than two years old found that ALL the infected plants showing symptoms had been from infected cuttings which still has old leaves showing VNPV symptoms. It suggests that infected cuttings were the main means of transmitting VNPV over long distances, for example, between islands in Tonga.

Most of the vanilla cuttings for plantations on Tongatapu had come from Vava'u in the past, so there is a high probability that the viruses were also unwittingly transmitted into new plantations because of the lack of quality control and disease knowledge.

Vanilla cured bean export is Tonga's most important export in terms of agriculture production. Prior to the discovery, identification and characterization of the VNPV, in 1988, the export of vanilla cured beans which was less than 20 tonnes of cured beans per year, was very low compared to the acreage on record. It was found during my epidemiology studies that many plantations were suffering from virus infection and new plantations were established with infected planting material. Many established old plantations had 80% of vanilla plants infected and showing symptoms.

It was a pleasant surprise that after the control measures were put in place and applied in the field, the vanilla cured bean export jumped to around 70 tonnes of cured beans in only a few years, in the late 1990s. All the symptomatic vanilla plants in the whole country were removed and replanted with vanilla cuttings from healthy vanilla. That is the single most effective way of beating the viruses.

FAO Stats did confirm that Vanilla production in Tonga had exceeded 200 tonnes of cured beans per year in the last decade (2008-2018). It does mean that production and export has increased exponentially after the VNPV was removed from the production equation. Tongan export of vanilla cured beans had been under 20 tonnes for 30-40 years, from 1960-1990....then it jumped to more than 180-200 tonnes per year in the period 1991-2019. The VNPV control measures were established by myself and MAFF advisory and extension staff during the period 1989-1990 after my return from Auckland University with the results of the research there which culminated in the award of the degree Master of Science in May, 1989. I ran a series of training workshops for MAFF Trainers, Extension Officers and Advisory Staff from 1989-1991 including radio programmes and field work/training. The Advisory Staff were tasked with removing all VNPV infected plants from all vanilla plantations in the Kingdom and showing Vanilla growers how to select the best cuttings for their new plantations.

We also produced a leaflet (Pone,S.P. and Pearson M.N., Koe Mahaki Vailasi 'o e Vanilla) for use by MAFF staff which has pictures showing diseased plants and advise on how to remove infected plants and replace them with healthy cuttings.

It does look like that the exponential growth of cured bean production only occurred in the two decades after the VNPV was removed from the vanilla plantation and growers fields.

In my last visit to Tonga in 2015, I collected pictures for my publications. I did a quick survey of tree crops like breadfruit, avocado and mangoes and important cash crops like vanilla and yams, I noticed that vanilla plantations did not have any obvious VNPV diseased plants but management has not improved. The plantations still look like the ones I surveyed during my work in 1986-1991. We have preached the gospel of management to improve disease control and production, since I started working with Mr Stephan Sorin on vanilla, but the old style planting is still used. It confirms my suggestion that the jump in cured bean production from less than 20 tonnes to more than 200 tonnes of cured beans is solely due to removal of the VNPV disease....and not due to any improvement in growers management style.

I want to put forward a suggestion to the Tongan Government that we should look at removing CyMV and ORSV from the field as well. We should produce CyMV and ORSV free planting vanilla cuttings and give them to vanilla growers…and also keep in mind the success that we had after removing VNPV.

It is a case now of getting rid of the CyMV and ORSV in order to get healthy plants for the industry, and further increase production. Although VNPV can be eradicated using only the symptoms as a identification tool, CyMV and ORSV need virus testing tools like ELISA and tissue culture techniques like meristem culture to produce virus free plants. Virus free vanilla plants are still available from regional Tissue Culture Laboratories like SPC and USP. That might be the easiest way of getting started.

Removal of CyMV and ORSV will push Tonga's Vanilla Industry past the 500 tonne mark, for example, based on the acreage being planted making it an important, crucial and vital step in Tonga's Vanilla Industry improvements . The logic is simple, healthier vanilla plants will produce better results. Tonga will be the first vanilla producing country in the world to do this....remove all viruses from its Vanilla plantations.

Chapter 14....

Taro Leaf Blight

Taro Leaf Blight (TLB) is suggested to have been introduced into the Pacific Islands by the Japanese Army during World War II. It was reported to have been present in many parts of Asia before World War II. For example, the first report of TLB was by Marian Raciborski in 1900 from Java, Indonesia.

Some publications point out that the Japanese Army was cultivating taro as part of the soldier's diet during World War II probably because it can grow very easily on the islands they occupied. Consequently, TLB was present only in islands which were occupied by the Japanese Army, or had contact with the Japanese Army. Those islands included Guam, Papua New Guinea, Solomon Is and Federated States of Micronesia.

The introduction of TLB into Samoa and American Samoa, in 1993, is a mystery, because the disease had not moved out of the original area of distribution since the end of World War II, a period of 40 years.

I was working for the Pacific Regional Agriculture Programme Tissue Culture Project 7, hosted by the University of the South Pacific's Institute for Research, Extension and Training in Agriculture (IRETA), from April 1992 to April 1993; based at the Alafua Campus, Apia, Samoa. The Samoan taro was excellent and the country exported $T10million worth to New Zealand, and other markets, every year. However, after the TLB epidemic in 1993, the country only exported $T160,000 to New Zealand and other markets. A reduction of 98.4% in taro exports.

The Samoans preferred the *Talo Niue* variety because it produces large corms and the flesh is firm. It also has a purplish color which adds to its popularity....the Manu Samoa, the National Rugby Team of Samoa, colour being blue, I suppose. As a comparison, *Talo Lau'ila* is the preferred variety of taro in Tonga. It has a white flesh which is softer than the *Talo Niue*. *Talo Niue* is also the main taro variety grown in Fiji. It is probably because *Talo Niue* responds well in wet conditions. *Talo Lau'ila* performs better in dry conditions, or islands with less water. That is probably why the Tongans prefer it, although *Talo Niue* is also grown in Tonga.

The two varieties can be easily differentiated. *Talo Lau'ila* has a very pronounced 'black spot' on the leaf surface, where the stalk meets the lamina. That is the meaning of *lau'ila* in Tongan, 'leaf with black spot'. It is just a pronounced 'base' of the leaf veins. It has a green stem/stalk and leaves and stalks tend to grow 'wider' rather than 'taller' so it does not grow as high/tall as the *Talo Niue*. The tallest *Talo Lau'ila* I have seen was only about 3-4 feet high.

The *Talo Niue* I have seen in Samoa grows to six feet or taller with dark purple stem/stalk and it has purplish, firm corm flesh.

I left Samoa to join the South Pacific Commission (SPC) at the end of April 1993. I was the new Plant Protection Officer, later Plant Protection Advisor, and Co-ordinator of the SPC Plant Protection Service, taking over from Robert Macfarlane who has been there the previous six and half years.

I was a bit shocked when the request from both Samoas came in July 1993 for assistance with the TLB epidemic. They were not sure what the disease was, which had destroyed more than 80 percent of the taro plantations, in both islands, by July 1993. It was obvious that the environmental and atmospheric conditions in the Samoas favored the spread of the disease in the months before.

The SPC Secretary General, Ati George Sokomanu sent our Agriculture Programme Manager, Seumanutafa Dr Malcolm Hazelman, the request from the Samoas. I found it with a note in my inwards tray. I immediately requested and allocated funds for my travel to both Samoas for a week to assess the situation and see how we can help.

I was driven around the island of 'Upolu by MAFF Officers, to inspect the damage. You can see the large dry spots on the taro leaf from the road. On the island of 'Upolu, the main road encircles the island, so it was possible to do the survey without getting off the vehicle. I wanted to know what the extend of the spread was. On some occasions where the damage is 100%, I wanted to get off and have a closer look. I was hoping to find an important prerequisite to disease severity like the presence of herbaceous weeds laden with aphids, in the case of the VNPV epidemic in Tonga. TLB is a wind and rain dispersed disease unlike the vector driven VNPV.

TLB sporangia can be splashed across from infected to disease taro plants during wind and rain showers, but only in short distances. Oospores are produced during unfavorable growing conditions and they can germinate when conditions are favorable. These Oospores can be carried on taro corms, and planting material, for long distances thus spreading the disease very quickly over long distances and between islands. It is unclear whether sporangia can be blown long distances and between islands by wind, which would explain the rapid widespread distribution of TLB in 1993.

Here's an excellent description of the symptom in Wikipedia;

Quote...

(*TLB*) Symptoms on leaves initially occur where water droplets accumulate and eventually form small, brown spots surrounded by (*yellow*) halos on the upper surface of leaves. These spots expand very quickly and form large brown lesions. The entire leaf can be destroyed within a few days of the initial appearance of symptoms under wet conditions. The undersides of leaves have spots that look water-soaked or gray, and as they expand, blight forms and the leaf is destroyed within a few days.Symptoms occur in a day/night pattern where water soaked areas expand during the night and then dry out during the day. As a result, additional water marks form leading to increasingly larger lesions. As the lesions expand, sporangia develop most actively at the margin of the lesion and progress to attack healthy tissue.

One characteristic feature found on leaves is the formation of bright orange droplets oozing out from above and below water soaked leaf surfaces. As a result, the droplets dry out during the day and become crusty. Another sign of *P. colocasiae* infection are masses of sporangia that form a white, powdery ring around the lesion. Symptoms on petioles includes gray to brownish black lesions that can occur anywhere on the petioles. Petioles become soft and may break as the pathogen destroys the host.

Symptoms on corms are often rubber-like and soft as well as having a light tan color. These symptoms occur rapidly and can arise anywhere on the corm and are often subtle in early stages. Decayed corm tissue appears brown and turns purplish in advanced stages of infection.

Lesions can also be formed by sporangia that are splashed by rain. The dead central area breaks and falls out as the lesion gets larger. The rate of spread for this disease is very high which results in a high percentage of yield loss.

Unquote...

During my July 1993 survey, in both Samoas, the disease was present in every village and causes so much damage that most plantations were wiped out, especially in the higher elevation, for some unexplainable reason. It could be due to more wind and rain in the hilly areas.

It does seem that the spread of the disease could not have been due to wind and rain alone. There may have been spread via planting material and sale of taro corms, which would explain the widespread distribution in such a short time, between villages and islands. Like the epidemic of VNPV in Tonga, ALL infected young plantations of less than 2 years old were from infected cuttings.

It is well known that Samoa has been exporting taro to American Samoa for decades. If the disease had started there they could have unwittingly introduced it to American Samoa.

It is unlikely that TLB was introduced from American Samoa because they have no taro export or planting material export to Samoa.

The widespread nature of the disease point to one fact, the disease may have been around for a few weeks or months, allowing planting material and taro corms to be transported all over the islands before the epidemic occurred when conditions were perfect, for fungal growth and it just exploded and destroyed 98 percent of taro export.

I had a meeting with the Director of Agriculture Tuisugaletaua Sofara Aveau, Ministry of Agriculture Officials, a FAO representative, USP representative and I think there was also a representative from USAID. The assistance that we can give to MAFF to allow it to plan and execute programmes to overcome the TLB problem were discussed.

I flew to Pagopago the next day to meet the Agriculture Officials there and I also had a meeting with the Governor of American Samoa. He understood the problem and was keen to see any solutions offered. What was obvious was that the taro plantings in Tutu'ila were all totally destroyed. In most taro plantations, only the stalks are still standing, all the lamina have rotted and fallen off.

I offered the meeting in Apia and also the Governor our assistance from SPC which I will get approval from the Secretary General first. We will bring together all the available expertise to discuss the TLB problem and propose solutions.

On my return to Suva, I sent my recommendations to our Management which include SG (Secretary General- Ati George Sokomanu), DP (Director of Programmes- Mafaitu'uga Va'asatia Poloma Komiti), DS (Director of Services- Mrs Fusi Taginavanua) and M/Agri (Manager Agriculture- Seumanutafa Dr Malcolm Hazelman). I recommended convening a meeting of ALL the known TLB experts in Apia at IRETA, USP, as a matter of urgency, to discuss the issues and find solutions.

I also proposed a budget which was approved and we had the First Taro Seminar on November 22-26, 1993 at IRETA, USP, Alafua Campus, Samoa. The meeting was attended by more than 100 invited international experts and country representatives, local experts and advisory staff. Country reports were given, expert papers presented and we got a very good, informative picture of TLB and its effect on taro in the Pacific Region and around the world. The meeting also passed some very important and vital resolutions to combat TLB and reduce its limiting ability on taro production in Samoa, American Samoa and the Pacific region.

I also proposed a meeting in 2 years time to review what has been achieved and improve what needs improvement.

The Second Taro Seminar was held at Lae University of Technology, Lae, Papua New Guinea on 26-30 June 1995. My motivation for taking the meeting to Papua New Guinea was the success of Dr Anton Ivancic, and Department of Agriculture Staff, with taro breeding against TLB there. I had visited the Lae Research Station of the PNG Department of Agriculture, on several occasions, and met with the Head of Research Mr Sim Sar, Plant Pathologist Pere Kokoa, as well as, Dr Ivancic to discuss their programme and view the various resistant varieties…and even taste them. I was convinced that we can solve the TLB problem in the Pacific simply by asking PNG to share their success with the rest of the Pacific. My feeling with my various discussions with Sim Sar and Director of Agriculture, Mr Ted Sitapai that it will be the best way. I had stopped over in Port Moresby on several occasions to discuss the TLB problem and other issues with Ted and other Agriculture Officials.

I also visited Dr Rob Harding, Dr Brendan Rodoni and Professor James Dale of the Queensland University of Technology (QUT) and discussed the project proposal with them. They were very supportive and even helped to write up the proposal.

I visited Dr Paul Ferrar at the Australian Centre for International Agriculture Research, Canberra to discuss the idea with him and he supported it. He approved the project for one million Australian dollars.

The idea was to clean all the resistant cultivars/varieties from PNG at QUT, ensuring that it does not have any virus or other microbe and mass produce in Tissue Culture and send them for Plant Quarantine observation then distribution is Samoa. In my view, we can solve the problem of TLB in one year, 1996. Dr R Harding, Dr B Rodoni and Professor James Dale will look after the project at QUT.

However, I had a feeling that there was opposition to my plan; 1. Somebody brought up the issue of *alomae and bobone* viruses and whether we can safely test for them 2. We need to discuss the project and get formal approval from the authorities in PNG. Although Mr Ted Sitapai, the Director, was supportive we have yet to get Ministerial and Government views.

I knew I will be migrating to New Zealand at the end of my contract in May 1996, SPC management has advertised all core budget positions as a 'restructuring exercise', including my job and although I reapplied there are no guarantees I will have a job in June 1996. I felt that job security is very important because expatriates like myself, and my family, who leave our country to come and work for the region, in Fiji, depend on continued employment. My wife and two little children who were 7 and 5 years old, at the time, depend on me being employed.

As a result of that uncertainty and other issues, I recommended to Dr Paul Ferrar that it may be best to deal directly with the Samoan Government and bypass SPC because I may not be there to carry out the required work for the recovery of TLB in American Samoa and Samoa.

It is really interesting that it took more than 20 years for the taro export of Samoa to recover. Our plan with Dr Rob Harding, Dr Brendan Rodoni and Professor James Dale would have allowed Samoa, and

American Samoa, to receive large numbers of tissue cultured TLB resistant taro varieties through the Queensland University of Technology beginning in 1996. Millions of TLB resistant tissue cultured taro plantlets can be given/distributed to Samoan growers by 1996/1997, for example. That will allow Samoa export of taro to New Zealand to resume by 1997/1998.

We can source the TLB resistant cultivars/varieties from PNG, Asia and other Pacific Islands and also develop reliable tests for *alomae and bobone* viruses which are present in the Solomon Is and PNG. I had developed a very reliable ELISA test for VNPV in Tonga, which helped remove it from the vanilla fields, and I was sure we can do the same for *alomae and bobone* viruses. ACIAR had approved $1 million for the project to start as a matter of urgency. I am sure that ACIAR would allocate more money after a satisfactory review. The QUT would have provided an excellent much needed service for Samoa, American Samoa and the other 20 Pacific Island members of

SPC using new high technology methods to improve our taro production, and disease control and ...later...other crops. We may even eradicate the *alomae and bobone* viruses in the process.

At the quoted export earning figure of $T10 million a year, the delay of 20 years before export resumed was not necessary, in my view. That is a loss of $T200 million for the Samoan taro growers.

I have seen the exported taro from Samoa, just 2 years ago (2017), in the supermarkets and fruit shops and even at the importer's warehouse, in Auckland, New Zealand. I must say that the operation looks very good. The packaging is good, the processing and product looks good and I have bought and tried the Samoan taro and they taste good. I had a talk with the importer and he said he buys about 160-180 containers of taro from Samoa per year. He says, the total Samoa taro export is about 200 containers a year or 4,000 tonnes. At a retail price of $4 per kilo that is a yearly income of $NZ16,000,000 generated by Samoan taro in the New Zealand market. If the

Samoan grower's are paid $NZ1 per kilo that is an income of $T7 million which is almost as high as the $T10 million Samoa was earning from taro export prior to TLB in 1992.

But did the Samoan growers have to wait 20 years for it? I say that we should have tried QUT and high technology which will enable us in the islands to do much, much more. QUT can provide high tech services, for our Pacific agriculture, that cannot be done in the Pacific Islands.

The Samoans were scared of the *alomae and bobone* viruses being introduced into Samoa through the QUT. That is the most quoted reason for not allowing QUT to do the work. Samoa had opted to try the Asian varieties and do their own TLB resistant breeding with international help, including ACIAR and SPC, and all the information given during the SPC Taro Seminars at IRETA and LUT (Lae University of Technology) which I had organized in 1993 and 1995.

Alomae and bobone viruses are not present in Asian countries.

Although SPC Land Resources has improved its agriculture technology a lot, judging from the information available on the SPC website, I still think that SPC should have some help from other organizations like QUT in terms of technology transfer and application. Why try to reinvent the wheel when others can do it for you for free? And also train your people in using the high technologies to improve agriculture production...as we have seen in the case of VNPV in Tonga?.

Chapter 15....

The Tongan Banana Industry

My first job as an Agriculture Officer/ Plant Pathologist for the MAFF Research Station in Tonga, in June 1985, was to help monitor the resistance of Banana Black Leaf Streak, caused by the fungus *Mycosparella fijiensis* conidial state *Paracercospora fijiensis*, to benomyl. I had just graduated with a Bachelor of Science from the University of Auckland on May 10th, 1985. I returned to Tonga on the 16th of June and started working on the 19th of June!

The Tongan Banana Industry had been slowly declining since its heyday in the late 1960s when Tonga exported 20,000 tonnes , or more, of bananas to New Zealand, annually. New Zealand had also started a $NZ5 million 'Banana Recovery Project' and was subsidizing much of the banana production to aid recovery. I was invited to join the Banana Technical Advisory Committee comprising MAFF experts and the New Zealand Banana Project staff.

Every week, I would drive in our new four wheel drive, double cab, courtesy of the Tonga-German Plant Protection Project to 10 pre-selected banana plantations on Tongatapu and collect samples of banana leaves. I selected the plantations on a geographical basis to cover the whole island. There was no pre-determining factor other than equal space between them.

Leaves collected show early signs of Banana Black Leaf Streak (BBLS) which I will 'dissect' and use in the Laboratory at Vaini Research Station for my tests for resistance against benomyl. Benomyl, as Benlate 50% WP (white powder), was the favorite chemical fungicide used in Tonga to spray against BBLS. It is a systemic fungicide which works well against all fungal crop disease under Tongan conditions.

Resistance against benomyl was reported from other banana producing countries, that it prompted the monitoring work in Tonga which I was responsible for.

The two tests for monitoring 'resistance' against benomyl;

These two methods for monitoring benomyl resistance has been used in many other countries.

Method 1.

Dissect the BBLS 'streaks', which were still small at about 0.5 cm long, from the leaf samples, and culture them on agar amended with 5ppm benomyl and non-amended agar, as a control. If there is hyphal growth after a few days, then I conclude that the fungus is resistant against benomyl. In most cases, there was no growth and I would advise the Banana Technical Advisory Committee of my findings.

(I had improved this method by using a pair of forceps rather than a scalpel to remove the 'streaks'. It reduces the work time by as much as 50% because it takes longer to cut out the streak with a scalpel).

Method 2.

Collect dead banana leaves/lamina and incubate them at high moisture for a few days. When the conidial growth on the banana leaf surface appears, the conidia (spores) are removed and used in the test by culturing in benomyl amended agar at 5ppm, 10ppm and non-amended agar. If the conidia grows in the amended agar, I conclude the fungus is resistant. In all cases the conidia does not grow in the amended agar but does grow in non-amended agar. It was interesting to note the distortion and stunting of BBLS hyphae in amended agar while they grow normally in non-amended agar.

The Tongatapu banana plantations were sprayed 2 weekly by a MAFF/Banana Project spray gang and the fungicides used were alternated to reduce fungal resistance to the chemicals especially benomyl.

My Senior colleague Mr Tevita Holo BSc, MSc was the Principal Plant Pathologist and he was a very good teacher and inspiration to the younger staff including myself. Tevita had designed several banana trials to check the efficacy of various fungicides, misting oil, as well as, nematicides. At about 10 total acres in size, they were the biggest trials at the Vaini Research Station at the time. The information collected were crucial to the continued reduction in banana production costs.

However, the Banana Industries of Tonga, Samoa and Fiji were all struggling at the time. The quality of the ripened bananas became more important because of increasing imports from Ecuador and Phillipines whose bananas were better quality in terms of appearance.

Methods of harvesting and packing bananas in Tonga were still primitive compared to Ecuador, for example. There was much bruising of the banana hands that show up as large black spots on the ripened bananas in Auckland. Post harvest rots caused by *Collectotrichun gloeosporioides* or anthracnose, as it was commonly called, were also important quality issues.

Despite a huge effort from ACIAR on selection of new banana varieties for export as well as the New Zealand Banana Project, the Banana Industry in the Pacific Islands was doomed. The continuing issues of quality and increasing production prices were critical to the survival of the industry.

The Banana Industry in Tonga finally succumbed in the late 1990s.

Some important notes.

1. The greatest cost to the Tongan Banana Industry was chemical control of BBLS.

Bananas were grown as mixed crops prior to 1964. After the beginning of the monoculture system which started in the mid-1960s, the production peaked at more than 20,000 tonnes in 1967-68. The inevitable introduction and spread of BBLS known as 'Black Sigatoka' because it was first found in the Sigatoka Valley, Fiji; occurred. BBLS spread like 'wild fire' in the large areas of banana monoculture, that it necessitates chemical control. By the early 1970s banana production had begun the slow decline until it was only about 500 tonnes, per year, in the mid-1980s.

2. The benefits of banana export to small holders

Although there was a lot of production costs, small holder farmers with small acreage of 4 acres or less made up the majority of growers. They were still able to make a significant income of $200 or more every two weeks when the Banana Boat visits. Compare that with the 1985 salary scale for Tongan Civil Servants. Bachelor degree holder's starting salary level was $163 a fortnight. You can appreciate the importance of that income to the small farmers. The larger growers were able to make much more every fortnight.

The crucial turning point, in my view, was when the subsidies from the New Zealand Banana Project ended. It took up much of the costs that farmers were lulled into a 'false comfort zone'. It was only when the subsidies ended that they realize that they cannot afford to pay for all the costs. I suspect that farmers did not want to pay so much for the production, and simply gave up.

ABOUT THE AUTHOR...

Semisi Pule also known as Semisi Pule Pone, uses Semisi Pone for short, was born in the island Kingdom of Tonga in the South Pacific in 1961. He attended Longolongo Primary School and Tonga High School in Tonga (1967-1979) then Mt Albert Grammar School and the University of Auckland (1980-1984) in New Zealand. He graduated with a Bachelor of Science in May 1985 and joined the Ministry of Agriculture, Fisheries and Forests in Tonga in June 1985 as an Agriculture Officer/Plant Pathologist.

He was awarded a scholarship by the German Government through GTZ, and the Tonga-German Plant Protection Project, to do a Master of Science programme doing research on the Vanilla viruses (1987-1988) at Auckland University. He graduated in may 1989 and continued his work with MAFF.

He was appointed to the position of Senior Plant Virologist for his work on Vanilla, Kava and Squash viruses in 1991.

His work in Tonga is published in many scientific papers and his series of books and ebooks on PLANT PROTECTION IN THE PACIFIC (Books 1-4).

In March 1992, he joined the PRAP Project 7 at IRETA, USP as a Fellow in Tissue Culture. His work there is published in the book PLANT PROTECTION IN THE PACIFIC 3, Tissue Culture.

He joined the South Pacific Commission Plant Protection Service in April 1993 as its Plant Protection Advisor and Co-ordinator of the Plant Protection Service (1993-1996). His work at the SPC PPS are published in his series of books and ebooks on PLANT PROTECTION IN THE PACIFIC.

He is now a writer with more than 200 books and ebooks in amazon.com and blurb.com.

He also operate a small contracts business.

I WISH TO THANK THESE PEOPLE AND ORGANIZATIONS...

New Zealand

My studies in New Zealand at Mt Albert Grammar School and the University of Auckland were made possible because these kind hearted people decided to help;

1. The most Reverend Taniela Takapautolo Moala and the Tongan Methodist Church of New Zealand who sponsored me and helped in every aspect of my stay in New Zealand (January, 1980 - June,1985).
2. Farmers Trading Company, Hobson St, Auckland CBD, who gave me a part-time Shop Assistant Job (1982 -1984).
3. Sachet Packaging of Penrose who gave me a summer part-time job, 1981.
4. Eden Park Maintenance Team who gave me a summer part-time job, 1984-1985.
5. Milano Restaurant, Parnell who gave me a part-time, evening job.

6. AUSA Job Search for getting me my part-time and summer jobs.

Kingdom of Tonga

1. The Hon. Lord Tuita, Acting Minister of Agriculture, Fisheries and Forests, June, 1985 who gave me a job as Agriculture Officer/Plant Pathologist with MAFF.
2. Mr Tevita Fa'oa Holo, Principal Plant Pathologist, Head of Research, MAFF for his scientific brilliance and discussions which helped.
3. Mr 'Ofa Fakalata, Principal Entomologist, Head of Research, MAFF for his humor and social skills.
4. Mrs Paelata Vi Nai, Lab & Field Technician for her hard work and helpful attitude.
5. Mr Soane Taula, Field Technician for his hard work.

Tonga-German Plant Protection Project

1. Dr Dirk Stechmann, Team Leader for his helpful discussions, attitude and assistance....and for the award of a Master of Science Scholarship from GTZ...and offer of a PhD Scholarship. (I actually asked Dirk I needed a break after the MSc but I did not know the PhD scholarships have to be used up before the end of the project...which effectively ruled me out to continue after a few years break! The Tonga German Plant Protection Project ended in a few years time...from 1989!).
2. Mr Konrad Engleberger, Team Leader for his help and assistance and skill assisting my scholarship management.

Samoa

1. The University of the South Pacific especially the Vice-Chancellor Professor William Pattie.
2. Dr Mary Taylor for her useful discussions and expertise in Tissue Culture.
3. Mr Muhhamad Umar, Institute for Research, extension and Training in Agriculture.
4. Mr Athur Nagatalevu

…for their administration assistance

Fiji

1. Mr Ati George Sokomanu, Secretary General (based in New Caledonia)
2. Mr Poloma Komiti, Director of Programmes (based in New Caledonia)
3. Ms Fusi Taginavanua, Director of Services (based in New Caledonia)
4. Dr Malcolm Hazelman, Manager, Agriculture Programme (based in Fiji)
5. Mr John Roache, Finance Manager (based in New Caledonia)
6. Mrs Keresi Morris, Personal Assistance (based in Fiji)

For all their administration assistance with my large number of projects!

A Special thank you....

Unlike most Tongan students who went to study overseas, on a Government Scholarship, I was a private student. My uncle, the most Reverend Taniela Moala of the New Zealand Tongan Methodist Church sponsored me to come and study in Auckland. First at 7th Form at Mt Albert Grammar School (1980) then Auckland University (1981-1984).
I stayed with my uncle's family and worked, part-time, during the term and also full time during the summer breaks, of every year, to pay for my university fees, books, bus fare, clothes, medicine, entertainment and so on.
I will always be thankful for the assistance of Rev Taniela Moala and Mrs Viena Moala and their children Nehasi, 'Iunisi, Siueni, Tu'utu'ulaki, Vai and Audrey but especially the oldest adopted son Mr Afe Sikaleti and his wife Mrs Fetu'u Sikaleti.

Their warmth and help, in every aspect of my stay in New Zealand, made everything possible.

Further Reading...

**All these prints and ebooks are
available from amazon.com**

1. Plant Protection in the Pacific (print)
2. Plant Protection in the Pacific 2 (print)
3. Plant Protection in the Pacific 3,
Tissue Culture (print)
4. Plant Protection in the Pacific 4 (print)
5.   Zucchini Yellow Mosaic Virus, on
Tongan Squash Pumkin (ebook)
6. Kava Dieback, recommended control
of the disease (ebook)
7. Vanilla Necrosis Potyvirus, control of
the virus in Tonga (ebook)
8. Vanilla Necrosis Potyvirus, second
edition (ebook)

www.ingramcontent.com/pod-product-compliance
Lightning Source LLC
Chambersburg PA
CBHW020155200326
41521CB00006B/385